ROUTLEDGE LIBRARY EDITIONS:
ORGANIZATIONS: THEORY & BEHAVIOUR

ORGANIZATIONAL CHOICE

ORGANIZATIONAL CHOICE

Capabilities of Groups At The Coal Face Under Changing Technologies

The Loss, Re-Discovery & Transformation of a Work Tradition

E. L. TRIST, G. W. HIGGIN, H. MURRAY AND A. B. POLLOCK

Volume 28

Routledge
Taylor & Francis Group
LONDON AND NEW YORK

First published in 1963

This edition first published in 2013
by Routledge
4 Park Square, Milton Park, Abingdon, Oxon OX14 4RN
605 Third Avenue, New York, NY 10017

Routledge is an imprint of the Taylor & Francis Group, an informa business

British Library Cataloguing in Publication Data
A catalogue record for this book is available from the British Library

ISBN: 978-0-415-65793-8 (Set)
eISBN: 978-0-203-38369-8 (Set)
ISBN: 978-0-415-82567-2 (Volume 28) (hbk)
eISBN: 978-0-203-43632-5 (Volume 28)

Publisher's Note
The publisher has gone to great lengths to ensure the quality of this reprint but points out that some imperfections in the original copies may be apparent.

Organizational Choice

CAPABILITIES OF GROUPS AT THE COAL FACE UNDER CHANGING TECHNOLOGIES

THE LOSS, RE-DISCOVERY & TRANSFORMATION OF A WORK TRADITION

E. L. TRIST
G. W. HIGGIN
H. MURRAY
A. B. POLLOCK

TAVISTOCK PUBLICATIONS

First published in 1963
by Tavistock Publications (1959) Limited
11 New Fetter Lane, London E.C.4
and printed in Great Britain
in 12 point Bembo by
C. Tinling & Co., Ltd., Liverpool, London, and Prescot

CONTENTS

INTRODUCTION

The preliminary studies towards this book began over ten years ago in the period, very different from the present, which followed the nationalization of the British coal industry. Expectation then was widespread that management-worker relations would improve and that productivity would increase simply through the change-over from private to public ownership. Such results, however, were not so easily forthcoming. Moreover, men continued to leave the pits though they were needed, and the incidence of stress illnesses at the coal face remained high (Halliday, 1949). Indeed, it was a medical source—the observations of a doctor practising in a colliery district in the interwar period (Dickson, 1936)—which provided some of the first evidence in favour of the general theory of organizational health and work effectiveness which underlies this book.

The view is advanced that the type of work organization conventionally associated with the main persisting form of partially mechanized longwall working contains identifiable socio-psychological features which lead to a number of substandard results: a rate of productivity below system potential; inflated face costs; poor management-labour relations; a low level of job satisfaction; and a high level of worker absence, whether voluntary or occasioned by sickness or accidents. Such a view could be rigorously tested only when, for this same technology, an alternative form of work organization became available, with features which would lead to a prediction of the opposite effects.

Though a number of forerunners of more limited application appeared in the early fifties, it was not until the mid-fifties that a widely applicable alternative emerged. It was the good fortune of the authors to have the opportunity of making a systematic and long-term study of this alternative—in comparison with the conventional organization—in the same seam, in the same neighbourhood, with the same type of equipment and the same type of men. Our findings are that, when the socio-psychological factors involved are thoroughly understood and taken into account, the alternative yields the improvements expected. When they are not, there may be losses rather than gains.

In mining, as in other industries, there is considerable resistance to recognizing that socio-psychological factors are in-built characteristics of work systems rather than additional—and possibly optional—features to do with 'human relations'. This attitude is one reason why the alternative work method has not been widely adopted.

The advent of mechanization has evoked from the industrial world a rather rigid adherence to a mechanistic theory of organization (Katz and Kahn, 1951). It is likely to be some time before an organic theory is as fundamentally accepted (Shepard & Blake, 1961). Yet, paradoxically, it is the further advance of technology which seems most calculated to bring this acceptance about (Woodward, 1958). Where coal face operations have become comprehensively mechanized, work groups are beginning to establish themselves which have wider autonomy, greater powers of self-regulation, and a fuller commitment to more holistic tasks than those associated with partially mechanized methods. Such characteristics parallel those of the emergent alternative at the level of partial mechanization discovered in this research. They are also the characteristics of pre-mechanized face groups. A tradition deeply embedded in the industry is being re-discovered and adapted to new circumstances.

An understanding of this tradition and the relevance of its application, appropriately transformed, is becoming even more important as mechanization of the industry proceeds—whereas in 1960 under 40 per cent of output was power loaded, by the end of 1961 it was over 50 per cent, and rising steadily.

It may be asked how the tradition came to be lost. The answer can only be that the loss of the more organic type of working group was difficult to avoid given the extent to which an ideology of extreme work breakdown accompanied the introduction of mass-production methods in industry generally. But now that the validity of this ideology has been called into question, much may be learnt concerning the optimum organization of working groups from an industry such as mining. Having always possessed latent traditions in an alternative direction, it has a creativeness which is bringing into existence models relevant in other contexts.

Of particular interest to the student of social process is the ability of quite large primary work groups of 40–50 members to act as self-regulating, self-developing social organisms able to maintain themselves in a steady state of high productivity throughout the entire period of their 'missions'.[1] These missions, which involved the daily management of a three-shift work cycle by the group itself, lasted for nearly two years—the 'life' of the coal faces concerned. At the end of this time the groups were still growing in their capacity to adapt to changes in their task environments and to satisfy the needs of their members. Autonomous groups of this size are not usually thought capable of succeeding with a task of this complexity or a mission of this duration. The degree of success actually attained varied widely among several such groups studied. Through a comparison, however, of the conditions attendant on these different degrees of success, some at least of the factors crucial for the effective functioning of large autonomous work groups have been identified.

The data on which this study is based could have been obtained only through the prolonged, patient, and intense collaboration of those in the industry, on both the Board and the Union side—at all levels. The relationship of the research team to the men and officials with whom they worked is described in some detail in the text as it is part of the method. We should like here to express our gratitude to them all for being prepared to give so much of their time and interest on the chance that an inquiry by a group of social scientists, about whose disciplines they could have had only the vaguest conceptions, might prove of some use to them as mining people.

The preliminary studies were financed through a grant to the Tavistock Institute of Human Relations by the Human Factors Panel (of the Government's then Committee on Industrial Productivity), which was administered by the Medical Research Council. Subsequent exploratory work was made possible through a grant for the general development of the Institute's research programme from the Rockefeller Foundation.

[1] For the use of 'mission' in this sense cf. Selznick (1957).

The main study, 1954–8, in Durham Division, was sponsored by the Department of Scientific and Industrial Research/Medical Research Council Joint Committee on Human Relations in Industry. It was financed at first from Counterpart Funds derived from United States Economic Aid and continued later from United Kingdom Funds.

In addition to the present authors, three former Tavistock staff members took part in the programme: Dr. A. T. M. Wilson, then Chairman of the Institute's Management Committee, was responsible for research policy in relation to the DSIR/MRC Committee and the Divisions of the National Coal Board with whom work was carried out. He contributed both to the fieldwork and to several source papers.

Mr. K. W. Bamforth collaborated in the original study. His experience as a former miner was a stringent criterion against which emergent hypotheses were tested. The pit in which he had worked provided the first example of an alternative to the prevailing work method.

During the main research period, Dr. P. G. Herbst carried out an independent study of a facework group in very great detail and developed a number of new concepts concerning primary group functioning. This study is offered as a supplementary monograph, under the title *Autonomous Group Functioning* (Herbst, 1962).

A second supplementary monograph is also under preparation by one of the present authors, Dr. H. Murray, on the quantitative assessment of composite performance. It has been impossible to do justice in this volume to the mass of material he collected and the systematic methods of treatment he worked out.

This book, which is an overall presentation, is based on a series of source papers by members of the research team. These are listed in the special bibliography and are available for reference through the Tavistock Institute of Human Relations or the Department of Scientific and Industrial Research. In each chapter reference is made to the particular source papers on which the account depends.

Professor Leon Festinger of Stanford University acted from time to time as an external consultant to the research team, visited the collieries in Durham where the main fieldwork was carried out, and gave invaluable criticism concerning the handling of data. During the preliminary studies Dr. F. E. Emery, then of the University of Melbourne, gave similar assistance.

The final manuscript was prepared by the first author—who has directed the Institute's field studies in the coal industry since their inception—while a Fellow, for the year 1960–61, at the Center for Advanced Study in the Behavioral Sciences, Stanford, California. It expands an earlier report made by Dr. Murray and himself in 1958 to the Joint DSIR/MRC Committee, which was presented to the industry.

PART I

PATTERN AND PROCESS

SECTIONS

ONE. *The Nature of the Project: Methods and Concepts*

TWO. *Traditional and Conventional Work Group Organization*

THREE. *Emerging Forms of Work Group Organization*

SECTION ONE

The Nature of the Project: Methods and Concepts

CHAPTERS

CHAPTER I

The Socio-Technical Approach

THE DEVELOPMENT OF THE CONCEPT

This presentation of research studies by the Tavistock Institute of Human Relations in a number of pits in North West Durham is concerned with the interaction of technological and social factors in industrial production systems—here represented by a variety of mining methods at differing levels of mechanization. The approach adopted, that of considering each production unit as a *socio-technical system*, originated in the first mining study carried out by the Institute (Trist and Bamforth, 1951). The usefulness of the concept having been demonstrated by subsequent work (Wilson and Trist, 1951; Trist, 1953), it has been further developed in two parallel Tavistock projects, one in the Indian textile industry (Rice, 1958), the other the present research. Wilson (1955) has noted that work on similar lines has developed independently in various countries and that similar findings have emerged (Walker & Guest, 1952; Westerlund, 1952; Touraine, 1955).

The propositions underlying the present studies may, following Trist and Bamforth, 1951, be stated as follows: '. . . the longwall method will be regarded as a technological system expressive of the prevailing outlook of mass-production engineering and as a social structure consisting of the occupational roles that have been institutionalized in its use. These interactive technological and sociological patterns will be assumed to exist as forces having psychological effects in the life-space of the faceworker, who must either take a role and perform a task in the system they compose or abandon his attempt to work at the coal face. His own contribution to the field of determinants arises from the nature and quality of the attitudes and relationships he develops in performing one of these tasks and in taking one of these roles. Together, the forces and their effects constitute the psycho-social whole which is the object of study.'

Rice, in *Productivity and Social Organization*, continues more generally: 'The concept of a production system as a socio-technical system designates a general field of study concerned with the interrelations of the technical and socio-psychological organization of industrial production systems. . . . The concept of a socio-technical system arose from the consideration that any production system requires both a technological organization—equipment and process layout—and a work organization relating to each other those who carry out the necessary tasks. The technological demands place limits on the type of work organization possible, but a work organization has social and psychological properties of its own that are independent of technology. . . . A socio-technical system must also satisfy the financial conditions of the industry of which it is a part. It must have economic validity. It has in fact social, technological and economic dimensions, all of which are interdependent but all of which have independent values of their own.'

It is, of course, the socio-psychological (the people) and the technological (the things) which are the substantive dimensions. The economic dimension measures the effectiveness with which human and technological resources are used to carry out the primary task (cf. Williams, 1950). The importance of the distinctiveness of territory has been discussed by Miller (1959). Emery & Trist (1960) have further shown that the socio-technical concept requires to be developed in terms of open rather than closed system theory, especially as regards the enterprise-environment relation and the elucidation of the conditions under which a steady state may be attained:

'Considering enterprises as "open socio-technical systems" helps to provide a more realistic picture of how they are both influenced by and able to act back on their environment. It points in particular to the various ways in which enterprises are enabled by their structural and functional characteristics ("system constants") to cope with the "lacks" and "gluts" in their available environment. Unlike mechanical and other inanimate systems they possess the property of "equi-finality"; they may achieve a steady state from differing initial conditions and in differing ways. Thus in coping by internal changes they are not limited to simple quantitative change and increased uniformity but may, and usually do, elaborate new structures and take on new functions. The cumulative effect of coping mainly by *internal* elaboration and differentiation is generally to make the system independent of an increasing range of the predictable fluctuations in its supplies and outlets. At the same time, however, this process ties down in specific ways more and more of its capital, skill and energies and renders it less able to cope with newly emergent and unpredicted changes that challenge the primary ends of the enterprise.' (Op. cit. p. 94.)

Inherent in the socio-technical approach is the notion that the attainment of optimum conditions in any one dimension does not necessarily result in a set of conditions optimum for the system as a whole. If the structures of the various dimensions are not consistent, *interference* will occur, leading to a state of disequilibrium, so that achievement of the overall goal will to some degree be endangered and in the limit made impossible.[1] The *optimization* of the whole tends to require a less than optimum state for each separate dimension.

FOCUSING ON THE SOCIO-PSYCHOLOGICAL SYSTEM

This approach does not imply that in all circumstances a detailed study of all three dimensions must be carried out. It does, however, underline the importance, when any aspect of a production system is examined, of taking into account the manner and extent of its interdependence with the other dimensions. In the present mining studies the research focus is the *socio-psychological* system. It is through the people who comprise this system that technological and economic changes are successfully or unsuccessfully implemented. For such changes to be effectively introduced, understanding of the latent as well as the manifest functioning of the socio-psychological system is necessary (Merton, 1949; Jaques, 1951; Blau, 1955).

The socio-psychological system may be studied at different

[1] Such dissonances between system characteristics may be similar to those described by Festinger (1957) in the field of cognition, though it is beyond the scope of this book to press such a comparison.

organizational levels in the coal as in other industries: at the level of the individual worker, the work group, the seam, the pit, the Area, the Division, or at the level of the National Coal Board itself, that is, the enterprise as a whole, when a very wide economic, political, and socio-cultural environment must be taken into account. The unit of study on which the present research is centred, however, is the *primary work group*. This is the smallest group whose membership carries out the whole set of activities constituting the unitary cycle of coal face operations. The boundaries of this social unit are defined in terms of the technological unit—the work cycle which it has to perform.[1] Just as the technical system of a coal face forms part of a larger system—the seam—in which it must be integrated for effective working, so does the primary work group—the cycle group—form part of a larger social system. The research is therefore concerned not only with the component work groups at the coal face which make up the cycle group but also with other individuals and groups in the seam population with whom they have immediate relations and who constitute the surrounding 'seam society'.[2]

At the level of the cycle group, the technological, economic, and socio-psychological dimensions differ in the degree to which they constrain modification of the system by the group. There is least freedom in the technological system for other than very minor modifications, decisions on the mining side rarely being taken below pit level and frequently involving higher management. In the economic dimension there is somewhat more, though still limited, opportunity for change, as in initiating local negotiations regarding the basis or amounts of payment. Strict account, however, must be taken of the framework of existing agreements, which may be seam 'prices' or local colliery settlements; and at an early stage any proposal has to be considered in terms of county and national agreements. It is the socio-psychological system which affords the greatest opportunity for either formal or informal change at the level of the cycle group—in such matters as altering the pattern of work group organization.

[1]cf. von Bertalanffy (1950) for the need to represent the mediating boundary conditions (here the technology) among the system constants in order to show how an open system achieves a steady state.

[2]cf. Lewin (1951) for the importance of adjacent systems both above and below the focal level.

It was not within our terms of reference to consider aspects of the economic system such as the capital, operating, maintenance, and wages costs, or the level of wages or piece-rate prices, as such. None the less, the form of the wages system has considerable bearing on the structure and functioning of the socio-psychological system and in this context is taken into account. Our principal concern is to examine that aspect of the socio-technical whole—the socio-psychological—within which the primary work group has relatively greater opportunity to develop various forms of work organization within imposed technological and economic limits. A set of concepts for describing the socio-psychological dimension is called for, which can be co-ordinated to concepts used in describing the technological.

THE RESEARCH OPPORTUNITY

Over a number of years pilot studies of a variety of mining methods had been made by the Institute in a number of coalfields. Because these studies could be made only as opportunity arose, and because the seam conditions, customs, practices, and attitudes differed in each locality, systematic and detailed comparison of the mining methods was scarcely feasible. Further progress required that these should be simultaneously available for study and that pits using them should be in one Area of a coalfield so as to minimize differences in background and tradition. The opportunity to undertake such studies in an older Area of the Durham coalfield was particularly welcome. In the collieries offered for study there existed—often in the same pit and all actively functioning in the present—a wide variety of mining systems ranging from traditional unmechanized working, through partially mechanized conventional methods, to more highly mechanized emergent systems. Since the faces concerned were in the same low seam, the geological structure of which was noted for its constancy, a comparative study of systems at different levels of mechanization was made possible under conditions more closely similar than any hitherto available.

In the most widespread of the conventional technologies in Durham there also existed two radically different forms of work organization, one of which had its roots in the earlier traditions of the coalfield, the other reflecting a form of organization more widespread in manufacturing industries. Comparison of alternative forms of work organization within the same technology therefore became feasible. The hypotheses emerging from the earlier Institute studies made the carrying out of such an 'experiment of opportunity' a matter of central scientific interest.

The co-existence in the present of a historically related range of mining methods, the growing importance of low seams as the higher were exhausted, and the increasing use of more highly mechanized methods in low seams also presented an unusual opportunity to observe, as they occurred, the socio-psychological aspects of technological change.

This book, which covers the period January 1955 to March 1958, presents a general account of the research findings together with a series of field experiments and case studies, full technical accounts of which have been given in the source papers. A description of the design of the research and the methods used is followed by presentation of the concepts developed for the appraisal of work systems. The main mining methods are then examined in these terms, proceeding from the simpler to the more complex and more highly mechanized. In the field experiments and case studies a comparison is made of the operational effectiveness of alternative forms of work organization at the most commonly found level of mechanization. Accounts are given of the social development of composite work groups under advantageous and disadvantageous conditions. Finally, we consider problems of changes in work organization both with and without accompanying technological change.

CHAPTER II

Design and Methods

One of the main problems to be overcome in communicating any work dealing with mining is that of terminology. Each coalfield has its own vocabulary for describing the systems, equipment, and processes of mining, and in the older coalfields, such as North West Durham, many pits use a number of entirely local expressions. Though the N.C.B. has attempted to rationalize the terminology used in the industry, in this account North West Durham usage is followed. So far as possible minimum use is made of specifically mining terms. A guide to mining terminology has, however, been prepared and is offered in Appendix I. Our interest is to develop a set of concepts which relate the technological and the socio-psychological systems. Since we shall be primarily concerned with longwall methods, illustrations of the concepts developed will refer to longwall organizaton.

SOCIO-TECHNICAL SYSTEMS IN MINING

The mining methods with which we are concerned have one characteristic in common—all are *cyclical* systems in which a sequence of operations has to be carried out for each 'web' of coal that is extracted. They are to be distinguished from more recently developed *continuous* methods in which operations are carried out simultaneously rather than successively so that a relatively steady flow of coal emanates from the face during all shifts. In cyclical systems, three phases of the production cycle may be identified:

(a) *preparation*, in which operations are directed towards making the coal more accessible and workable
(b) *getting*, in which the coal is loaded and transported away from the face

(c) *advancing*, in which roof supports, gateway haulage roads, and conveyor equipment are advanced.

Depending on the type of technology, these phases comprise a varying number of operations of different duration which only to a limited extent overlap in time.

The character of socio-technical systems is related to level of mechanization. As in other industries, mechanization in mining has proceeded from the introduction of power driven tools which aid the miner towards the introduction of more complex machines which he services and controls. Systems vary not only in level of mechanization—the proportion of effort coming from non-human sources—but in the evenness and comprehensiveness of the mechanization of their component parts.

In the simplest form of traditional mining each work place comprises a small separate coal face and is occupied by one miner at a time, although the place itself may be shared by two or three men, each of whom works on a different shift. The coal is won by hand picks and removed from the face in tubs. Various patterns of laying out the faces for working a seam are to be found. These methods—bord and pillar, rib and stall, gateway and stepwise longwall—are similar except for layout and can, so far as work organization is concerned, be treated as one type, and referred to as *single place working*.

The advent of *face conveyors* led to the introduction of *longwall conveyor working*. On these straight faces, 80–120 yds long in this part of Durham, coal was originally won, as in single places, with hand picks and then filled (shovelled) on to a face conveyor running along the now continuous coal face and discharging into tubs in the gate. The later introduction of pneumatic picks, both in single places and on hewing longwalls, raised the level of mechanization a further step. Another variant of longwall technology which raises it still further is the use of electrical coal cutters and shotfiring to prepare the coal for loading by hand. Of these two variants of partially mechanized longwall conveyor mining, cutting longwalls are the most widespread in Durham and throughout Great Britain, whereas hewing longwalls are peculiar to Durham and one or two other areas. This predominance of cutting faces arises because most of the coal is rather hard and at the same time most of the seams are fairly level. On the European continent hewing faces are more common as much of the coal is both relatively soft and in heavily faulted areas. In the United States continuing availability of thick seams near the surface has led to the mechanization of bord and pillar layouts in shallow mines and drifts.

Conventional longwall organization has developed on the principle of 'one man—one job', but an alternative form has emerged on some hewing and cutting faces which has its origins in the single place tradition. This is known as *composite longwall* working in which there is no rigid division of labour as on conventional faces.

Mechanization of conventional longwalls began in higher seams with the preparation and, to a lesser extent, with the getting phase. Relatively simple methods for power-loading prepared coal are now also becoming more widespread in low seams, and more advanced methods have been introduced which combine the preparation and getting phases of the production cycle. At the present time further mechanization is taking place by the installation of power-assisted methods for advancing face conveyors, roof supports, and gateways.

THE EXPLORATORY PHASE

The initial conception of the research, based on outline information about the methods of working in the pits offered for study, was to compare, within a single seam, earlier traditional methods with conventional longwall working in both its hewing and cutting forms, and with forms of this latter containing elements of higher mechanization. Divisional and Area executives felt that the results obtained from increasing mechanization were far from always up to expectation and wondered how far the reasons might lie in the socio-psychological field.

One of the main starting assumptions was that a single seam of coal would be sufficiently similar at different pits to make possible direct comparisons of the results obtained by different methods of mining. It was found that, despite its reputation for general consistency, conditions in the selected seam—the Manley[1]—were not physically identical in the way hoped for. Certain geological

[1] The names given to coal seams are fictitious.

variations were present in each pit, which, though slight in themselves, had been decisive in settling the method of mining. There were also differences in the seam systems in which the various face units existed. These were to some extent technological, as for example the nature of the haulage facilities; but there were more general differences such as phase of development, scale of operation, and degree of homogeneity-heterogeneity in methods of facework.While such factors added to the difficulties of making direct comparisons between face units, their early elucidation substantially increased the reality of research plans. It became even clearer than had been anticipated that the cycle group of each face must be studied in relation to seam organization, for what was being done with the seam as a whole from the mining point of view greatly affected the psychological climate—and technological conditions—at particular faces.

During the exploratory phase other parts of the designated pits were visited to gain, for comparative purposes, familiarity with conditions elsewhere than in the selected seam. It was realized in this way that it did not so much matter whether a face was in one seam or another so long as the project was kept within the kind of low seam conditions generally prevalent in the Area, which the seam originally selected typified. This is because in longwall working seam height affects face length and this, in turn, the size of the working group. In higher seams visited during previous studies faces had been much longer and groups altogether larger than those now found. This makes the longwall system under high seam conditions a rather different world as regards group relations from that encountered in the present study—and even more difficult.

The general aim was restated as the comparison from a socio-psychological viewpoint of certain methods of mining under low seam conditions. While residual variations in geology and in operational background might prevent the simple direct comparisons of performance originally envisaged, the selected seam offered the best available example of what happened to work organization and group relations under low seam conditions. The development of valid methods of comparing performance became a problem to be resolved during the research. The methods developed are stated in Chapter XII.

DESIGN

As the study progressed, certain modifications were introduced into the original programme of pit visits so that a better design could be obtained. Work was undertaken at five pits, which between them provided three examples each of single places, hewing longwalls, cutting longwalls, and more highly mechanized faces. In one of the pits both conventional and composite forms of longwall organization were available for study.

The overall design involved the intensive case study, both qualitative and quantitative, over an extended period of time of the structure and functioning of the social system associated with at least one example of each of the main mining methods. Less intensive studies were made of other examples of each system but no attempt was made to conduct a 'sample survey', for it is not in this way that one can discover how a system works or changes over time. The period of work immediately following the exploratory phase was directed towards the elucidation of a set of concepts which would permit comparison of the different systems, and identification of those aspects of each which merited fuller consideration.

The research was conceived as falling into four main sections:

(a) A background study of the types of group associated with unmechanized pre-longwall mining—single place methods of working—represented by bord and pillar and rib and stall layouts.

(b) Comparison of conventional and composite methods of longwall organization on:
  (i) hewing longwalls
  (ii) cutting longwalls.

(c) Problems associated with the conversion of faces from pre-longwall and conventional organization to composite methods:
  (i) from bord and pillar to composite hewing longwall
  (ii) from composite cutting shortwall to composite cutting longwall
  (iii) from conventional to composite cutting longwall.

(d) The effects on cutting longwall organization of further mechanization:

(i) single task machines, e.g. flight-loader, scraper-packer
(ii) multiple-task machines, e.g. multi-jib cutter, Haarman scraper-peeler, tension-chain scraper.

SANCTIONING THE PROGRAMME

Following a request in August 1954, to the Chairman of the Durham Division, N.C.B., for an informal discussion concerning the granting of research facilities, a meeting was held at the Divisional Headquarters at which agreement was reached that work should go forward in the Division. A particular Area[1] of the Division was selected for study and a general research plan drawn up. A further meeting was held in November 1954, at which sanction for the research was given both by the Chairman of the Division, and by the General Secretary, National Union of Mineworkers, Durham Area.[2] The Institute then met the Area General Manager, his staff and pit managers to discuss in detail the proposed research methods and programme.

The initial procedure followed at every colliery was as follows. The colliery manager met the secretary and other officials of the local N.U.M. branch (lodge) to outline the purpose of the research and to inquire whether the lodge would agree to fieldwork being carried out at the pit. A full meeting of the Colliery Consultative Committee was then held at which the Institute outlined project objectives, research methods to be used, and procedures to be followed in reporting back results. It was agreed that detailed reports on work at particular pits would be submitted for comment to the Colliery Consultative Committee before being passed, on the N.C.B. side, to the Area General Manager and the Divisional Chairman and, on the N.U.M. side, to the General Secretary, N.U.M. Durham Area. It was further agreed that any more general report would be seen by these latter executives, before being discussed as regards general publication with the National Coal Board in London. This procedure would ensure that all reports would be accurate as regards matters of fact and

[1] For purposes of both technical planning and general management collieries are grouped geographically into Areas, which contain about 20 collieries, although this number varies greatly according to the size and dispersion of the collieries involved.

[2] On the Union side an Area is equivalent to a N.C.B. Division.

non-prejudicial to the interests of the Board and the Union, while the views expressed and the interpretations made would be those of an independent research team.

METHODS

Fieldwork was conducted by observation at the coal face on all shifts and by interviews with key informants at all levels from workmen to managers. Underground visits lasted from four to seven hours. The first objective was to gain an understanding of the technology and of the customs and traditions. On initial visits to a face we were introduced by both a senior management representative and a union official. During the subsequent visits, as far as possible, a different guide was provided on each occasion, sometimes from the union, sometimes from management. In later phases, when the research team had become at home with the geography, safety rules, and customs of the different pits, members were given the normal written authority of the pit manager to travel unescorted.

Although the men on the faces were aware of the general nature of the research, the first period constituted a phase of testing out the research team. We were questioned closely about the reasons for the research, the level of sanctioning, and the source of financial support. Once the face teams accepted that the research was not being carried out on behalf of either management or union, but was an independent scientific inquiry and that the results would in the first place be reported back to those directly concerned, discussion of 'private' rather than 'public' attitudes and opinions (Katz and Schanck, 1938) about systems of work became possible. From what the teams told us during these early visits of their attitudes towards 'official' research, effective rapport would not have been established had any attempt been made to take notes at the coal face. The practice was therefore followed of making tape recordings from immediate recall following each underground visit.

One research team member assumed responsibility for maintaining relationships with the team of each working face, observer reliability being ensured by independent visits from other members and by concurrent visits of pairs of fieldworkers. The validity of facework observation was assessed by the method of predicting the activities at one part of the face from a knowledge of what was going on in another, and then checking the predictions. Because of restricted visibility and mobility under low seam conditions, this is the only feasible way of obtaining a valid account of a system in operation. As observations became more elaborate and a greater understanding of the system was achieved, the provisional descriptions and conclusions were discussed informally with the men concerned to ascertain whether a true account had been obtained. By following this procedure of repeated delayed playback the research team became aware of on-going changes in work organization. Events were uncovered in this way which had been regarded as too insignificant to mention, but which were of key importance for understanding the system.

Although all our informants were interviewed at the coal face, a number were also seen, individually or in groups, in rooms at the miners' lodge. The men were unfamiliar with this kind of situation, and such methods proved unproductive. They were discontinued in favour of less formal contact in men's homes and in local clubs and pubs.

Full access was given by management to the standard pit records dealing with production, attendance, sickness, absenteeism, etc., and to the various statutory reports prepared by colliery officials. For some aspects of the research, use was made of specially devised record forms which were completed by deputies and team captains.

Although the general pattern of the research became established at an early date, the unpredictability both of conditions at the coal face and of management-lodge relationships made it impossible to follow a programme predetermined in detail. The research team had often to await the resolution of circumstances beyond their control and be on the alert for opportunities not envisaged when plans were drawn up. Particularly was this so in pits in which technological changes were introduced, or in which the system of work organization was being altered.

Without the complete co-operation of both management and union the research would not have been possible. In fieldwork of this type a research team is more than usually dependent on the help and guidance of those with whom they work. The increasingly open manner in which problems were discussed and the readiness of face teams to throw light on matters which would not in other circumstances have been disclosed were essential conditions on which obtaining the primary data depended.

CHAPTER III

The Appraisal of Socio-Technical Systems in Mining[1]

PRIMARY TASK

The concept which integrates the technological, economic, and socio-psychological aspects of a production system is the *primary task*—the work it has to perform (Bion, 1950; Rice, 1958). Work in this sense is the key transaction which relates an operating group to its environment and allows it to maintain the steady state (Emery & Trist, 1960). The production system with which we are concerned is the coal *face system*, although as indicated in the previous chapter this must be studied in the context of the larger *seam system* of which it is a part. The primary task of a face system engaged in three-shift longwall working is the daily completion of a production cycle under all the given conditions that prevail.

THE UNDERGROUND SITUATION

The underground situation can vary greatly from one face to another and from one type of system to another, but common to all is the absence of fixed and consistent conditions in the physical environment. The complex of factors affecting work at the coal face is of the kind that would confront a factory if productive machinery had to be moved and re-set every day; if every operator had to contend with constant minor changes in the material he was working on; and, at the same time, look to keeping the walls and roof of his work area supported because they were imminently liable to collapse; if all supplies had to be

[1] Source papers: 18, 20, 22, 23.

brought in and products removed through two narrow passages; and if, despite the absence of uniform working conditions, supervisors could visit the operators only occasionally throughout one shift. Unlike the factory situation, where a high degree of control can be exercised over the production process since working conditions can be maintained in a passive and constant state, in the underground situation the threat of instability from the environment makes the production task much more liable to disorganization. Under the dangers, stresses, and difficulties of the underground situation, certain qualities, evolved from the experience of successive generations and characteristic of traditional mining systems, are especially appropriate for the organization of work groups:[1]

(a) acceptance of responsibility for the entire cycle of operations
(b) recognition of the interdependence of one man or group on another for effective progress of the cycle
(c) self-regulation by the whole team and its constituent groups.

How far a work group is capable of such *responsible autonomy* and is able to adapt itself in correspondence with changing conditions indicates the extent to which its social structure is appropriate to the demands of the underground situation. The concept of responsible autonomy introduced by Trist and Bamforth in their original study of longwall in relation to traditional systems is intended to summarize these requisite[2] characteristics.

ACTIVITY STRUCTURE

Analysis of the activity structure in these terms is basic to an understanding of work group organization—the way in which those who carry out the necessary tasks are related to each other. Although the technology places limits on the kind of work organization possible, it does not uniquely determine its form, which may be analysed in terms of the five following aspects, each of which is more fully considered in subsequent paragraphs:

(a) the quality of the *work roles* to which each system gives rise through the division of labour

[1] cf. Gouldner (1954).
[2] For the term 'requisite organization' *vide* Brown (1960).

(b) the kinds of *task group*—the groups who together carry out given operations at the coal face and share a common pay-note
(c) the *work culture*—customs, traditions, and attitudes—which governs how these groups are built up and conduct themselves
(d) the nature of *inter-group relations* between task groups making up the face team
(e) the *managing system* through which the work of all faces in the seam is supervised, supported, and co-ordinated.

WORK ROLES

In analysing any mining system, or in comparing one with another, the first consideration is to make a systematic examination of work roles. By work roles are meant the jobs which people do every day and with which they become identified—cutters, fillers, pullers, stonemen, etc. In thinking of themselves as such they gradually take on certain common characteristics and may be said to acquire the *character* of their role. In an industry such as mining, the role with which a man is identified becomes a way of life. Work roles vary both within and between mining systems, in their nature, their quality, the demands they make upon their occupants, the satisfactions they afford, and the degree and type of stress to which they expose those who carry them out.

A work role—what a man does, where, when, and with whom—is primarily determined by the formal division or allocation of the tasks constituting the cycle of operations among the men who form the cycle group. A distinction must, however, be made between the formal or specified work role (the 'model') and the actual role which develops under a particular set of operating conditions. This distinction between formal and informal functioning applies also to other aspects of the social system. Recognition of such differences may not only indicate the efficiency of system functioning but also point the direction in which explicit technological or social change may profitably develop to achieve a better fitting together of the different aspects of the system as a whole.

A formal work role usually constitutes a main task together

with such sub- and ancillary tasks as are associated with it. Since in certain systems the shifts on which main tasks are carried out are fixed, it is meaningful to talk of *task-shift roles*. According to the system of organization, tasks may or may not be specific to particular roles. The range of tasks and shifts comprising a work role may be narrow or wide depending on whether there is formal *rotation of shifts and tasks*. The *task range* of a role may also be increased by disorganization, as when the work a man normally does becomes unavailable because of cycle breakdown and he is required to undertake activities properly belonging to another role. The delineation of work roles is, therefore, to some extent a function of the period of time over which the role content is observed.

The task content of work roles must also be examined in relation to the level of skill involved, a consideration of importance where questions of degree of specialization or interchangeability arise. How far the activities of a role comprise a self-completing whole task which occupies a full shift or how far the man has to fill out part of his shift by undertaking other activities has bearing on the extent to which he can experience satisfaction and identify himself with his role. In work systems where there is a large number of different work roles, some, because of the position they occupy in the cycle, may gain in status, power, and reward at the expense of others, to a degree which is disproportionate to any real difference in skill and effort.

All roles at the coal face are stressful, but the pattern varies with the different roles. The physical effort required, the liability to interference from factors beyond the control of the facework group, the cruciality of the tasks for cycle progress, the monotony or variety of tasks, the permanency of the roles and the shifts on which they are carried out, are all factors which contribute to the *stress pattern*. When conditions become bad, certain roles are more exposed than others so that undue stress falls on those concerned. Casualties arise if there can be no relief. The pattern of absence, accidents, and sickness is, therefore, relevant in an examination of work role stress.

TASK GROUPS

The whole team of men responsible for the operations of a particular unit is the *cycle group*. This is sometimes referred to as the 'face' or 'panel' team. In any seam, the cycle groups, together with the other piece and datal workers in the seam, constitute the *seam population*, which includes the younger men who are aspiring to facework, as well as older men who have retired to jobs away from the coal face. Depending on the system of working and the form of organization, the cycle group may contain a number of distinct sub-groups. Where the cycle of operations is spread over more than one shift, there will be different *shift groups*. If a production unit has two faces—a double unit—there may be two *face groups*.

A group of men carrying out a particular operation in a particular location on the coal face is referred to as an *activity group*. These groups may vary in size—according to the particular activity or activities carried out—and in the kinds of relationships between the men constituting the group. According to the system of work organization, the membership may be permanent or may vary, systematically or otherwise. Those activity groups which share a common paynote, and in which membership is permanent, are referred to as *task groups*.

Task groups, which are the basic units in a study of facework organization, vary in size and differ according to the kind of relationships the men have with each other in the work situation. In *identical role* groups all concerned are supposed to do the same amount of the same task and work more or less independently of each other. In *reciprocal role* groups the interdependent component activities of a main task are shared out among two or more persons who work together in order to complete it. There are also task groups of one, *isolate roles*, in which a man carries out a main task alone. The first step in the analysis of the characteristics of each task group is to determine the extent to which the structure of the task places limits on the kinds of relationship that are possible between the men in the group—whether work may be carried out independently or in sub-groups of two or three; whether the level of skill or effort required is the same for all members of the group; whether or not stress arising from the task is likely to affect all members equally, etc.

WORK CULTURE

For a fuller understanding of the behaviour of facework groups—whether cycle, shift, face, activity, or task—it is necessary to examine the pit and seam *culture*—the customs, traditions, and attitudes which regulate how men achieve membership of the various groups and conduct themselves as group members. In the underground situation of high risk, self-dependence and the good use of discretion are necessary. It follows, as was pointed out at the beginning of this chapter, that to a considerable degree face groups must be autonomous and self-regulating. This can only come about if groups develop customs and traditions for regulating their behaviour and relationships which are internal to themselves and binding by force of the authority of the group itself. The psychological climate of a group and the kinds of relationship it has with other groups involved in completing the same primary task is to a considerable degree determined by the way in which the groups are built up. It is important to determine the route through which men achieve membership of particular groups, the permanency of membership, and the route by which men leave. To do this an examination must be made not only of the various facework groups, but of the wider seam population. The earlier history of the seam has to be examined in order to gain an understanding of the way in which some of the on-going pit customs and practices have evolved.

INTER-GROUP RELATIONS

In longwall working especially, the various task groups need some system of inter-group relations to enable them to co-operate successfully in the overall task of cycle completion. An appraisal must, therefore, be made of the way in which task groups are related to each other and the extent to which their activities and attitudes facilitate or hinder completion of the overall goal. The *degree of segregation* of the various task groups comprising the cycle group, their number, and work relatedness to each other determine the basic pattern of inter-group relations. Within task groups there may be a number of activity groups whose membership may or may not be permanent. Role allocation may be to varying degrees flexible. At one extreme are fixed roles, at the other systematic patterns of rotation; in between are *ad hoc* exchanges of jobs. The extent to which all members of a cycle group share a common experience must be examined in relation to the way each constituent group carries out its work, since this can affect the conditions with which later groups have to cope.

The technological interdependence of activities is such that task groups are to varying degrees dependent on preceding groups. A situation of this kind tends to give rise to differences in status and power—according to the relative independence of a group's work and its cruciality for cycle completion. Where differences in status and power arise they may be reflected in differences in the level of earnings. It is in this connection that account must be taken of the way in which the payment system operates. The earnings of a group may be dependent to varying degrees on completion of the overall cycle task, on completion of its own main task, on the efficiency with which preceding groups carry out their tasks, and on the amount of unpredictable interference with their work arising from causes beyond the control of the cycle group. When the earnings of one group are too greatly dependent on the skill and attitude to its work of another, conflicts may arise and a state of tension develop between successive groups. As a protective measure against loss of earnings caused by the inability to complete main tasks regularly, groups may also, wittingly or unwittingly, go into collusion to carry out work in ways which, though economically mutually advantageous to the task groups concerned, may militate against the efficient completion of the production cycle.

THE MANAGING SYSTEM

Because main task groups share work places and equipment on successive shifts, overall co-ordination and continuity are required for smooth running of the *face system*. This is provided by the managing or governing system—which includes the total means adopted to maintain the boundary conditions of given 'operating systems'.[1] These must be examined not only at face level but at the level of the seam. The *seam system* includes the face systems together with their common *service system* which covers everything supporting the productive operations of faces—transport facilities for coal and supplies, communications, manpower reserves, repair work, development work, etc. The formal managing system of the seam embraces the deputy, overman, and undermanager—officials who are external to the face team; but some co-ordinating and regulating functions may reside within the face team and its constituent groups. These may sometimes be carried more or less explicitly by 'team captains'.

Important for a differentiation of systems of work organization is the extent to which co-ordination of task groups is *internal* or *external*—is carried out by the cycle group itself or effected by management external to the face team. Specific activity and task groups may be internally self-regulating without accepting responsibility for co-ordinating themselves as a shift group. Shift co-ordination may be provided entirely by the deputy who, when his shift overlaps others, also provides continuity between them. It is, therefore, necessary to identify the level at which responsibility is taken for co-ordinating the cycle group as a whole and the means by which this is done.

Since at any time the progress of operations on a face may be affected by what is happening in other parts of the seam system, each face 'shift supervisor'—the deputy—must be related to the seam 'shift supervisor'—the overman—whose responsibilities vary according to which shift he is on and the character of the seam system. It is at the level of the undermanager that the two facets of seam management—co-ordination of the cycle of operations on each face and co-ordination of the seam system on all shifts—become the responsibility of one person. An examination of the managing system at each level in the seam must explore the way in which information is generated, received, and transmitted, the kinds of decisions that have to be taken, the means by which they are implemented, and the nature of the stresses and strains to which officials are exposed. At the deputy and overman levels in particular, which are usually filled by promotion of qualified men from the face, conflicts may be experienced between the pressures

[1] Rice and Trist, 1952; Rice, 1958.

of tradition and custom on the one hand and the technological and economic demands of higher management on the other. The nature of these conflicts and the habitual methods of coping with them must be ascertained. Finally an attempt must be made to assess, in terms indicated in the earlier paragraphs of this chapter, the extent to which the managing system functions so as to enable the production units to carry out their primary task under the ever-present difficulties of the underground situation.

SECTION TWO

Traditional and Conventional Work Group Organization

CHAPTERS

CHAPTER IV

The Single Place Tradition[1]

THE TECHNICAL EQUILIBRIUM

Bord and pillar and rib and stall layouts are still found in many of the low seam pits of North West Durham where residual areas of coal have to be worked or where the seam is approaching the outcrop and the type of cover is unsuitable for long faces. These layouts may be subsumed under the general heading of single place working. In this system of mining interlacing roadways are driven at right angles into the seam, leaving small square or rectangular pillars of coal, 30-50 yds in length, which are then wholly or partly extracted. A coal face 6-11 yds in length—called a 'place'—is worked during any particular shift by one, or at the most two, miners who hew the coal with pneumatic picks. If the coal is hard, shotfiring may be used. The coal is then shovelled by hand into half-ton tubs which have been pushed up to the coal face on rails by another miner known as a 'putter'. As the coal is extracted the roof is supported by timbering. If the seam is low, stone has to be removed from below the floor by digging or blasting, in order to make height for the tubway. The stone is then stowed into the space from which the coal has been removed, the roof settling on to the packs so made. Places are normally worked in blocks—called 'flats'—consisting of 12 or 13 'bord' places to one 'winning' place, which makes the heading.

Each separate work place is a small, self-contained coal-producing unit, in relation to which all facework operations are carried out independently. If adverse geological conditions slow or halt the work in one particular place, this does not affect neighbouring places. As equipment is simple—hand or pneumatic picks and shovels—little time is lost through mechanical breakdown. Coal

[1] Source papers: 15, 18, 20.

comes away from the face intermittently in small amounts—a tub at a time—the tubs being gathered from the places by hand or pony and assembled into 'sets' on a 'landing', which serves a flat, to proceed to the shaft bottom on a simple haulage system. Since small amounts of coal are coming away all the time, the pressure on the haulage is relatively constant and relatively light. Even if a certain amount of hold-up should occur in the supply of empty tubs to the face, it does not always interfere with production, despite the annoyance it causes, since hewn coal may lie for a time without interfering with facework operations. A general characteristic of single place working is that the service system which supports productive operations (haulage, flow of supplies, manpower reserves, etc.) and the face system match and balance each other.[1]

The dispersion of production over so many units slows the tempo, but a slow tempo is the easier to maintain. Very little organization is required and the seam as a whole becomes virtually self-regulating, for although there is a cycle within each work place, the sum of the intermittent individual contributions imparts a non-cyclic character to the whole. This is maintained through whatever shifts are worked, so that in a three-shift pit there is continuity of production throughout the 24 hours. Since the flow is not intense, the winding capacity of the shaft need not be great, nor arrangements elaborate for handling coal on the surface. These facilities can be fully used all the time, so that the greatest economies come from three-shift working with an even spread of personnel over each shift. This is the basic model of the system, though one-shift and two-shift pits occurred among those visited. In one of these, certain tasks such as timber-drawing (formerly done by the deputies) were carried out on a separate nightshift by special men.

Several concepts and traditions regarding work roles and group relations have emerged from single place working in North West Durham which are of the utmost importance for an understanding of more complex mining systems. The most important are the ideas of the composite, self-supervising workman, the self-selecting 'marrow' group, and membership in the seam unit involved in 'cavilling'.

[1]This may be regarded as a socio-technical equivalent of 'structural balance' (Heider, 1946, 1958; Cartwright & Harary, 1956).

THE COMPOSITE WORK ROLE

The cycle of operations in single place working is a simple one. It begins with breaking-in to the coal face and continues with hewing and loading the coal into tubs, while roof supports are set at intervals prescribed by regulations. When the face has been advanced a distance agreed among the group who occupy the work place, the stone in the floor is taken up and built into packs, and rails are then laid to allow the tubs to approach the coal face. This cycle of operations is not governed by a time schedule. Work proceeds in a continuous manner, each successive shift taking up the cycle at the point left by the previous shift. The flexibility of work pace so allowed has special advantages in the underground situation for, when bad conditions are encountered, the extraction process in a series of places can proceed unevenly in correspondence with the uneven distribution of these bad conditions.

A single place worker must, therefore, possess the necessary range of skills to carry out all face tasks, for sooner or later he will be required to do them all. He is a 'complete miner'—the collier—who supervises himself and is the person directly responsible for production. The *composite work role* which derives from the technological requirements of single place working has established the tradition and reality of faceworker autonomy.

THE MARROW GROUP

The primary work group—the *marrow*[1] *group*—is comprised of men who share the same paynote through working the same place either on the same or different shifts. In winning places where two men are on together there will be six in the group in a three-shift pit, and three for bord places where only one man is on at a time, or two if there are only two shifts. These groups select themselves, the men choosing marrows or mates of the same standard in work performance as themselves. Account is taken of an individual's capacity for physical effort, his skill as a workman, the standards of performance he sets for himself, his known pattern of attendance, his age, etc. Because men of like capacity tend to work together, the earnings of marrow groups vary widely, even for the same

[1] Pronounced mærə.

work place. Men were found earning scarcely more than the minimum wage; others were earning between two and three times as much. These extremes were partly the result of differences in conditions and partly of differences in ability and stamina.

The process of self-selection gives the group a sanction which can be either implicitly or explicitly applied in order to maintain its standards. A man failing to meet the group's standards will be warned informally by his marrows; if improvement does not occur he will be told to look for others at the next quarterly cavil. If the man is unsuccessful in finding them, he may be placed by management in a group with others in a similar position to himself, where each man is on his own rather than on a shared paynote; or he may find himself on datal work, which is less remunerative than facework. The small group, capable of responsible autonomy and able to vary its work pace in correspondence with changing conditions, is a type of social structure ideally adapted to the underground situation.

The existence of marrow groups has given the Durham faceworker great experience in handling relations in the small group. It has also given rise to the widespread belief that only small groups can work together successfully—because in large groups the range of individual differences would be too great for a common paynote to be equitably shared; either trouble would arise inside the group, or the group would restrict its output to that of the slowest member, while using its collective strength to bargain for a higher price—when trouble would arise with management.

SEAM STATUS AND CAVILLING

The seam is a unit with sociological no less than technological reality. It is a formal social group which confers rights and demands obligations. When accepted as a member, a man is said to have been granted *seam status*. This gives him a right to a work place before anyone lacking this status and he cannot be moved arbitrarily to another seam while there is work in his own. A man in fact joins a seam rather than a pit, becoming known, for example, as a Manley or a Bramwell man. He graduates through a series of roles to the coal face in his own seam, and, when he is older and leaves the face for less arduous work, it is into the backbye areas of his own seam that he goes. Recruits are first employed as datal hands on the seam haulage, moving on to work as putters, where they acquire knowledge of facework operations and become known to the various marrow groups. When vacancies arise they may be invited to join an existing group or form a set of their own.

Further rules concern the equitable sharing of work places so that every group has an equal chance of working in good and bad conditions. This is known as *cavilling*, a unique feature of the Durham and Northumberland coalfields. All work places, except development headings which are let as 'bargains', are pooled every quarter and drawn by lot in a formal ceremony at which both management and lodge are present. This system is highly adaptive to single place working, since there are a large number of places and wide differences between them, which directly affect earnings. In practice, cavilling does not always ensure equal sharing of the good and bad. What cavilling does ensure is a randomness of allocation that provides a safeguard against favouritism and victimization. This is the aspect which has appealed to the Durham miner, encouraging him in an earlier period to adapt for his own use a folk custom concerned with obtaining justice at the hands of chance. The institution and its results are accepted by management and men alike, although difficulties do arise over its procedures and application. The cavilling rules of a pit form a kind of case law which has been built up over the years. The continuance and meticulous enforcement of these rules are insisted on by the union.

These customs and rules produce a seam group which is a highly organized and stable population. This organization and stability do not stem from the management; they arise from formalization of the customs of the working group of which the lodge is the guardian.

MANAGEMENT CHARACTERISTICS

In the pits included in the present study, it has been traditional for workers and management to deal with each other through the lodge to a far greater extent than in most industries. The executive system, especially at junior levels, is correspondingly undeveloped. At the old type of pit there was, indeed, no need for more than a rudimentary executive system, since the self-supervisory character of the work roles and the slowness of the production tempo deriving from the absence of mechanization made the task of co-ordination relatively simple. On the other hand, the variations in working conditions at any one time and the constant changes in these conditions created the need for an amount and frequency of wage negotiation uncommonly great by ordinary industrial standards. Moreover, this negotiation contained a strong local element in view of the idiosyncrasies of different pits. These are conditions under which one would expect an elaborate trade union organization to emerge—with a strong local basis. The co-existence of a well developed trade union structure with an undeveloped management structure is a central feature of the social system of collieries as these have been fashioned by the earlier and simpler forms of mining.

The wages system that has emerged under these circumstances is based on the idea of the *price agreement*. This is a negotiated price for a specified task—the amount of coal got, or the amount of stonework necessary. These prices vary according to such conditions as the nature of the coal or stone. Where conditions become abnormal 'consideration' is paid by management. Such a simple piece-rate system with allowances for special circumstances matches the needs of single place working where there is a direct relationship between physical effort and productivity. Through its contractual emphasis it has led to a tradition of *management through the wages system* rather than through the direct executive control of the working group, which has remained autonomous. The means through which this form of management has been exercised is the relationship between the colliery manager and the trade union lodge, represented by its committee and especially by its president and secretary. It is management by negotiation between two independently constituted 'powers' who are yet mutually dependent, and whose relative strengths have varied considerably over the course of time. A complex system of agreements and procedures has grown up. So long as both sides abide by these, the primary task of producing coal can legitimately proceed; but usually present is the need somewhere to reach a fresh agreement in view of the changing character of undergound work; and there is always some anxiety as to which side will come off best in the next round. For any particular group 'everything goes right if the price is right, and everything goes wrong until it is right'.

Leaving aside formal wage negotiations and problems at colliery level concerning mining methods and policy, the function of day-to-day management in an operating seam of single place units is to provide and maintain the conditions which will best allow self-regulating faces to operate. Towards the single place worker the *deputy* stands in a service rather than an executive relation for, by law, the primary responsibility of the deputy in charge of a single place flat is with safety regulations; thereafter with the maintenance of supplies to the faceworkers. The deputy also carries out the duties of shotfirer, visiting the various places in his district to fire stone shots in bottom caunch if this is being taken, or in the coal itself if this is hard. It is at the next level, that of the *overman*, where the day-to-day regulation of the payment system takes place, together with the general co-ordination of seam operations on each shift. The distance of the overman from the faceworker is already considerable. He does not often visit a man in his place. Even greater is the distance of the *undermanager:* his overall statutory responsibility is for the area covered by a complete underground ventilation system, and this may include more than one seam.

CLOSENESS TO SYSTEM POTENTIAL

The cost of stonework for the large number of gateways required and the adverse effects of dispersion on the deployment of mechanized equipment have made single place working uneconomic in thin seams. Mechanization of single places is possible in shallow high seams, but few such seams are left and installations made after the war on the basis of American experience have restricted application. The important lesson from the present study of traditional systems is that, within limits of what is technologically simple, it is possible to realize system potentiality. Single place systems are fully adapted both socially and technically to the conditions of unmechanized working. They are simple and homogeneous, and face requirements and service facilities balance each other so that there is a minimum of interference. They are easy to manage, as at the level of face and seam alike they are largely self-regulating. There is only one work role—of high quality—since task and cycle group are the same, while the seam population deploys itself according to the rules of cavilling which are observed as an unquestionable tradition.

The only way in which productivity can be increased in these circumstances is by directly influencing the performance of the individual. Some of our older informants explained how in the years of economic depression bad relations arose with management from the pressure put on faceworkers to increase production. Price agreements were manipulated so that a 'living wage' could be achieved only by maintaining a level of output which placed men under unnatural strain. Fights were not infrequent over who should have use of tubs, which were often in short supply. One or two tubs extra on a man's note might mean that his wages at the end of the week would be marginally sufficient instead of short of his minimum needs.

Several teams in the pits visited had above average outputs and were undoubtedly already at their limit—sometimes beyond it. This was noticeable with some of the younger men in process of establishing homes and raising families. These men went all-out for big pay packets but bore considerable anxiety about 'burning themselves out'. Fear over contracting pneumoconiosis was excessive though not without a reality basis. Older men, past their peak, were worried about their ability to make the minimum wage and hence to find marrows. Isolation in the work place at this phase of life encouraged hypochondriacal attitudes of a depressive kind. Shift isolation, except when a man is visited by his putter or deputy, is the most disadvantageous aspect of single place working. It is, of course, absent in winning places which are larger and where at least two men are on at a time. To have two or more men either on at a time or on overlapping shifts was indeed the rule when seams were higher and a place could support a larger marrow group.

Special methods might have raised the level of the average groups to a slight extent, but on the whole the work pace adopted was the natural one for those concerned. More training might have been undertaken with below average groups, where lack of know-how contributed to poor results. To have raised the productivity, however, of any of the districts visited would have been a most difficult undertaking, without radical change of the entire system. In the flats studied in detail it was said that an extra tub per day from each place would have yielded an increase of some 7 per cent. No one, however, thought such a target capable of realization, considering that some men were already fully productive and bearing in mind the ages, ability, and stamina of others.

The production data for the Manley seam in the selected pits quoted to the research team at the beginning of the inquiry gave an average face o.m.s. of 2·5 tons for the rib and stall layouts and 2·4 tons for the bord and pillar, while the longwall conveyor faces yielded an average of 2·75 tons for cutting and 2·3 tons for hewing. As stated earlier, these results do not permit unqualified comparison even though the seam is the same and notably regular. Their general similarity, however, is meaningful, if not their differences, for Division had expected that the more highly mechanized conveyor faces would have returned a productivity beyond the reach of the single places. If these latter were working close to the limit of their potentiality, could the same be said of the conventional longwalls? If not, what was the nature of the difficulties preventing attainment of better results, and how far did they lie in the socio-psychological dimension? The next two chapters will attempt to answer these questions, which will be explored further in the case studies presented in Chapters XIII and XIV.

CHAPTER V

Conventional Longwall Working: its Emergence and Variants[1]

With the introduction at the beginning of this century of 'scraper' and, later, 'belt' conveyors, it became possible to extend the length of the working face. In North West Durham these straight longwalls are conventionally 80–100 yds in length. As distinct from the honeycomb of short faces contained in a bord and pillar flat, a longwall district has one continuous face, connected by end gates to a trunk road. A *single unit* longwall has a main roadway or 'gate' to the face, known as the mothergate, and a secondary road at the opposite end of the face, known as the tailgate. If the district is laid out as a *double unit* to obtain greater concentration, there are two faces, one on either side of the mothergate, each having its own tailgate and making up a panel of 160–200 yds overall. Each face is serviced by a conveyor, which loads the coal down one of the gates—the central one in a double unit—whence it is transported by another conveyor to the trunk road, where tubs or further conveyors take over.

THE FACE CONVEYOR AND CYCLE DOMINANCE

The traditional method of winning coal had for generations been some form of single place working. Not until the last fifty years has there been any fundamental change in this system, which has taken in drilling and shotfiring and later the pneumatic pick, while work roles and work organization have remained the same. It was the face conveyor that brought the first major change. Its introduction allowed the development of the modern longwall as distinct from gateway and step-wise longwalls,

[1] Source papers: 2, 3, 6, 15, 16.

which were no more than groups of single places laid out in series.

The longwall made possible by the face conveyor has a compelling economic advantage in that the proportion of stonework (to make gate roads) in relation to extraction area is substantially reduced. The exhaustion of the more accessible seams of thick coal and the need to work thinner seams increased the importance of the coal-stone ratio. There is also the question of extraction at greater depth where the lateral effects of pressure often crush short pillars, and longwall faces are preferred even under high seam conditions. For these two reasons longwall working spread rapidly in the inter-war period.

With the extension in the size of the coal face a fundamental change ensued in the technological organization of facework. Though single place working is cyclic in that the three main processes involved in winning coal (preparation, getting, and advancing) are present, they are not formally separated as regards work roles and shifts. With the longwall the scale of operations becomes such that these processes need distinct groups each working a whole shift, and the cyclical character of facework acquires central importance. The right balance must be found between manpower and task size and every effort made to complete on schedule the operation belonging to a particular shift.

EARLY LONGWALLS

The early forms of longwall were simple extensions of single place working, the techniques of preparation and getting remaining the same. The model for these is the *hewing longwall* in which the cycle has only two phases. In the first, which occupies two of the three shifts, preparation and getting are done together; in the second, advancing is carried out. With the introduction of the coal cutter, the cycle is broken into three phases as exemplified in the orthodox *cutting longwall*, with preparation on the first shift, getting on the second, and advancing on the third, in conjunction with, or overlapping, stonework. This situation makes cycle dominance even greater. Three phases, now undertaken by separated task groups on different shifts, increase problems of regulation and co-ordination. Moreover, both an upper and a lower limit to the amount of advance, and so of work done in all tasks, is fixed by the length of the cutting jib. With hewing, variation in advance can be used to meet varying conditions. With cutting, all work has to conform to the task set by the cutter, irrespective of conditions, unless the length of the jib is altered, which involves a major change. Rigidity in the technological system has replaced the fluid conditions of the single place, still to some extent preserved on hewing longwalls.

Hewing longwalls persist in areas of soft coal. In areas of hard coal, hewing, even with pneumatic picks, becomes too arduous to be efficient and cutting is preferred where conditions are otherwise suitable. On the European continent, hewing longwalls are still the most widespread. They are also common in some parts of Britain, among them North West Durham, where a single unit, 80 yds long, is worked by 12 *hewers* using pneumatic picks. On the first shift, half the hewers 'break in' to the coal and fill off as much as they can; on the second, the rest advance the face an agreed distance. On the third, while the face conveyor is being moved up and the roof supports are advanced by one set of men —the *pullers*—others—the *stonemen*—are enlarging the gateways by removing stone from the floor and roof and packing it in to the space from which the coal has been taken (the goaf).

THE CUTTING LONGWALL

The most widespread conventional longwall technology in Britain is, however, the cutting longwall. The layout of these faces is similar to hewing longwalls. On the first, or cutting, shift the coal is undercut to a depth of 4 to 6 feet with an electric coal cutter manned by two men, known as *cuttermen*, with another—the *scuffler*—clearing out the undercut. Holes having already been made by the *driller* before undercutting, the *shot-firer* (an official) then breaks down the coal by explosives. On the second, or filling, shift it is shovelled on to the face conveyor by seven men (on an 80-yd face) called *fillers*—from their having filled tubs before conveyors were introduced. As the coal is filled off, roof supports are set—timber or steel straps held up by timber or steel props, the more modern forms of which are adjustable. The coal at the head of the mothergate is taken by a hewer. On the third, or pulling and stonework, shift the conveyor and supports are advanced and the gateways enlarged as in the hewing longwall. This cycle of operations—cutting, filling, pulling and stonework—is the pattern followed in Durham. In some parts of the country the order of the cycle is different—pulling and stonework following cutting, with filling last.

The layout of a typical double unit is shown in *Figure 1*. *Figure 2* gives a section across the coal face, and *Figure 3* a section along the mothergate showing the positions of the tension box for the gate conveyor and the gearheads for the face conveyors in relation to the two caunches. All diagrams describe the state of affairs at the end of the filling shift. A more detailed presentation of the component activities of the cycle is made later in *Table 5* (p. 80), and the nature of their sequential dependence is set out in *Figure 4* (p. 82).

THE ORGANIZATIONAL BREAK WITH THE SINGLE PLACE TRADITION

Longwall conveyor mining, especially as this has come conventionally to be organized, represents a sharp break with single place working and the question arises of how far earlier traditions can be adapted. Several problems require examination: the persistence of customs—such as cavilling—inappropriate in the new situation; the importation of ideas about organizing mechanization from manufacturing industry—such as maximum job breakdown and work role specialization—inappropriate at the coal face; and failure to make constructive use of certain features of the older tradition—such as the marrow group of composite workmen.

In both cutting and hewing variants, activities are performed in a fixed sequence which constitutes a *formal work cycle* spread over three shifts throughout the 24 hours. If this timetable is not kept, production is lost, a situation entirely different from the informal cycle of the single place, which is non-cyclic at seam level. Coaling is concentrated into one or two shifts while the existence of the longwall face concentrates it spatially. This greater intensity of production over a limited period makes far more rigorous demands on the haulage and creates a need for a much higher level of co-ordination. The earlier concept of a relatively simple self-

FIGURE I

PLAN OF DOUBLE-UNIT LONGWALL AT END OF FILLING SHIFT

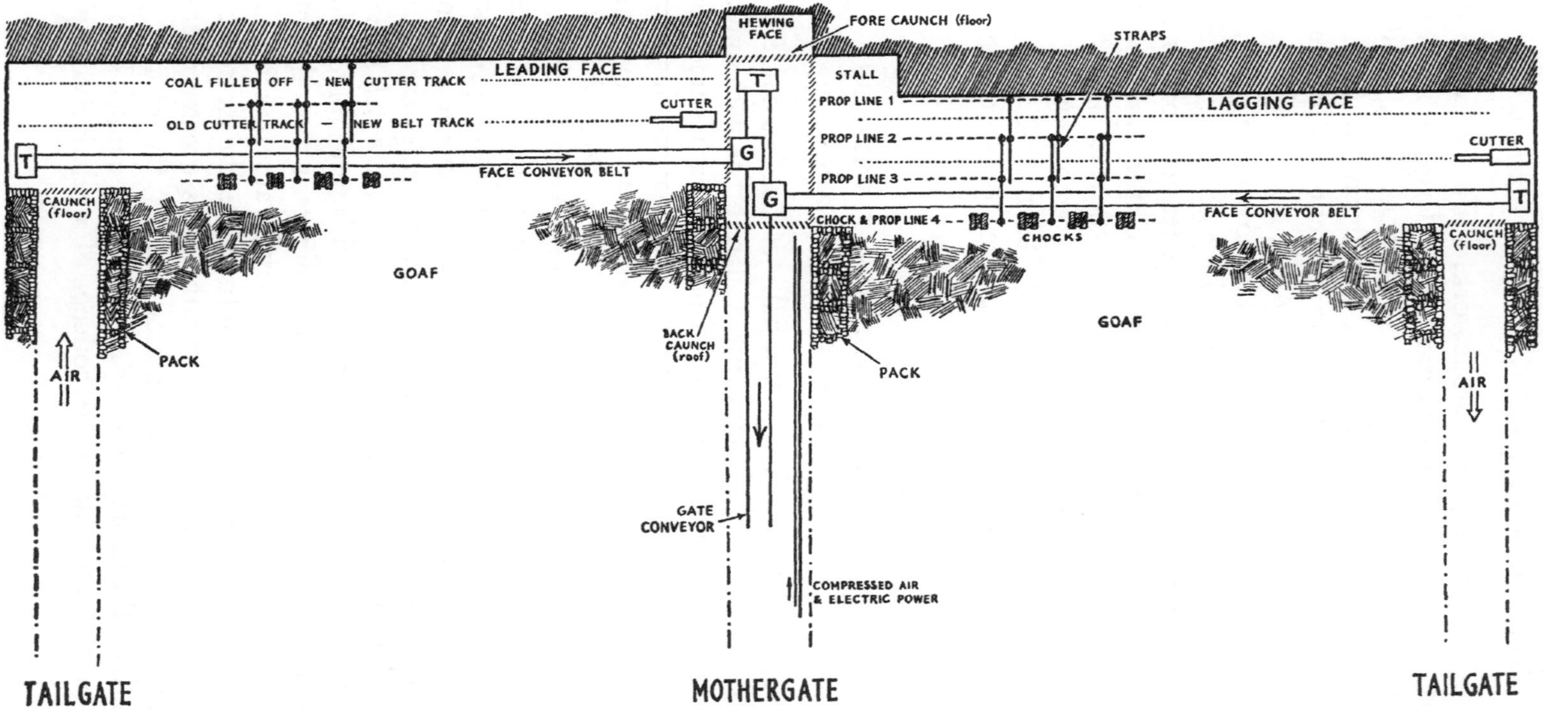

Key: G *Gearhead* T *Tension Box*

(not to scale)

FIGURE 2

SECTION ACROSS COAL FACE AT END OF FILLING SHIFT (not to scale)

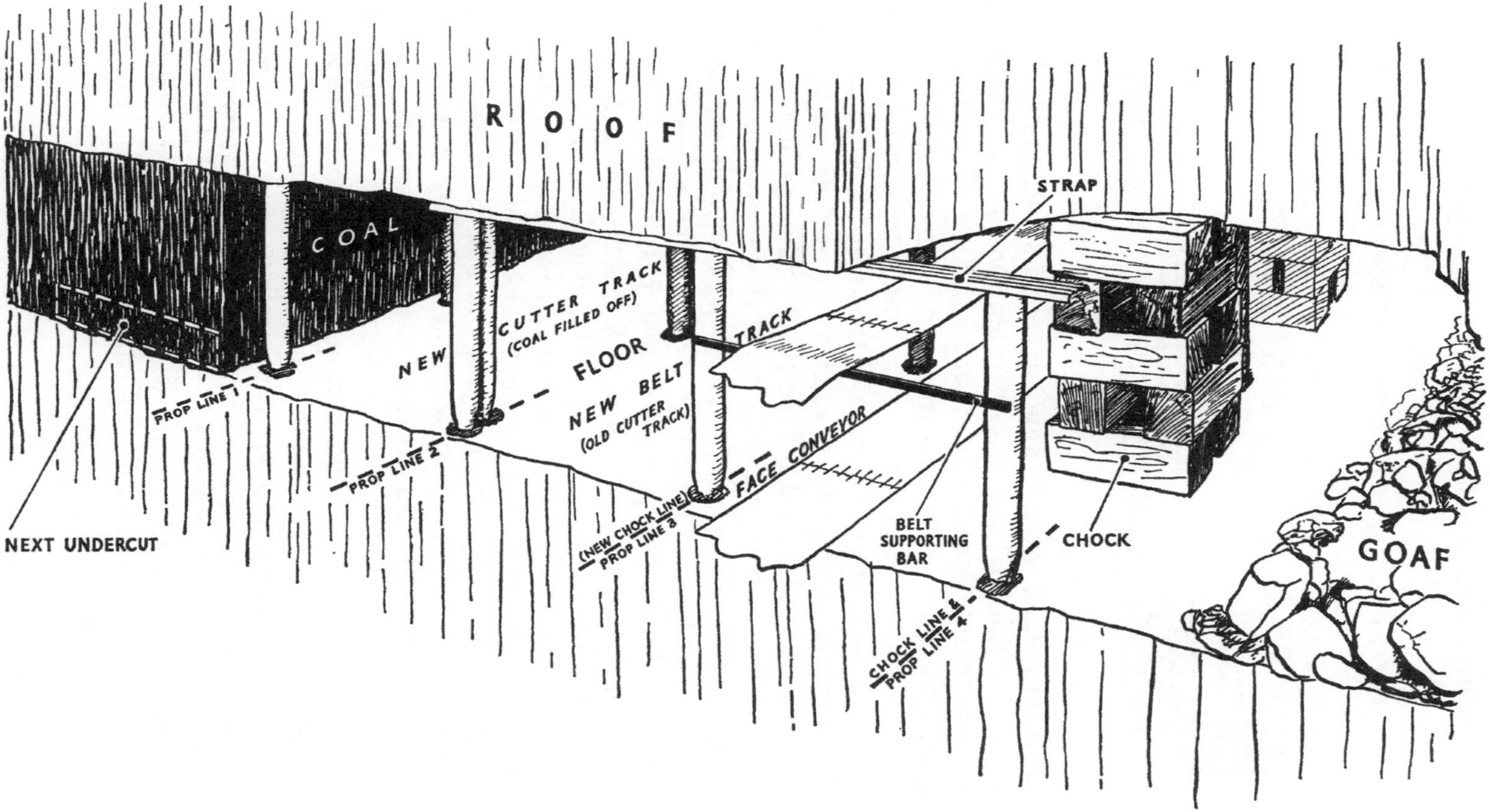

FIGURE 3

SECTION ALONG MOTHERGATE AT END OF FILLING SHIFT

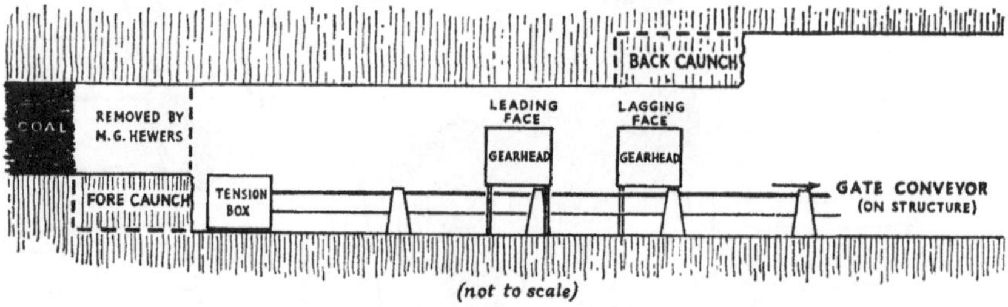

(*not to scale*)

regulating seam system cannot meet these demands; the carry-over of traditional supervisory roles, which have remained the same in title and largely also in function, has inhibited the development of a more appropriate managing system.

The unitary character of facework is replaced by a formal *division of labour* so that a series of specialized task roles, related to the sequence of activities, is carried out by different people. The way these are organized and the effect of this method of organization on the relations and attitudes of those concerned is the main subject of this chapter. Conventional longwall organization is characterized by the general conception: *one man—one task*. Each segregated task group has its own personnel, customs, price agreement, and paynote and is very largely bounded by its own concerns (cf. *Figure* 5, p. 83). No overall social organization exists to bind these segregated task groups together in common pursuit of the primary task of completing the cycle.

The single task group, carrying out one main task of the face cycle, such as filling or pulling, and extending its range of activities to any sub- and ancillary tasks that may fall within its domain—but no further—is the basic element in conventional longwall organization and is common alike to the hewing and cutting forms. It is into these single task groups, which, apart from certain exceptions, are self-selecting—and into these groups alone—that the marrow relationship is carried forward.[1] Previously the marrow relationship had encompassed the entire cycle group, which in single place working was the same as the task group. An understanding of the internal structure, external

[1] In other coalfields men are often assigned.

relations, customs, and psychological climate of these single task groups, in terms such as those introduced in Chapter III, is essential for understanding the conventional longwall.

THE OPPORTUNITY FOR ORGANIZATIONAL CHOICE

Given a technology of the kind that emerged, some division of labour became inevitable. The character of the various work roles in themselves, however, is not such as necessarily to compel the existence of the type of single task group which has become conventional. This would be so only if the recognized longwall face skills, such as cutting, scuffling, drilling, hewing, filling, pulling, and stonework, were sufficiently complex that the average faceworker could not, in a reasonably short time, be expected to become qualified in more than one. Under these circumstances their elaboration into separate occupations would be justifiable. The facts are otherwise.

In all work at the coal face two distinct tasks are simultaneously present. The first comprises the activities that belong to the production cycle. These, however, are always to some extent carried out on the background of a second activity arising from the need to contend with interferences emanating from the underground situation. Ability to contend with this second or background task comprises the common fund of underground skill shared alike by all experienced faceworkers. This common skill is of a higher order than that required simply to carry out, as such, any of the operations belonging to the production cycle. For these, initial training is short, several weeks or a few months, and it is now official policy for a man to be trained in more than one of the recognized face skills—preferably three. But the specifically mining skill of contending with undergound conditions, and of maintaining a high level of performance when difficulties arise, is developed only after a number of years at the face. A work system adapted to the underground situation must build this experience into its organization. Otherwise, it will fail to engage the faceworker to the limit of his capabilities and, indeed, restrict his performance.

The evidence to be presented in the rest of this chapter, and in Chapter XIII, suggests that conventional longwall organization acts in just such a restrictive fashion. Yet this form of organization is not compelled by the technology, which implies that an alternative is possible. The nature of this alternative is discussed in relation to its roots in single place working in Chapters VIII and XIV, and in Section Four.

DIFFERENCES BETWEEN HEWING AND CUTTING FACES

There are, however, certain differences between hewing and cutting faces which must be considered before conventional task groups are examined in detail. Two sets of conventional faces, in seams worked respectively by hewing and cutting, were visited repeatedly throughout the research. A number of others were visited over shorter periods and the present account is based on this experience as a whole, together with that gained in earlier studies in other coalfields.

The differences between hewing and cutting faces, as regards formal organization and manning, may be seen by comparing *Tables 1* and *2* (pp. 50 and 51), which represent the models of the two systems on which everyday practice was based in the pits included in the study. So that the comparison may be complete, the model for single place working has been set up in the same terms in *Table 3* (p. 52).

The contrast is self-evident between the unitary cycle of the single place system with its one work role and one task group, which constitutes the entirety of the small cycle group involved, and either of the longwall systems with their phased work cycles, differentiated work roles, and multiplicity of task groups making up relatively large cycle aggregates. These aggregates would be larger still were seams higher and faces longer. Under such conditions overall cycle groups of twice these numbers are common. Equally evident is the greater complexity of the cutting over the hewing longwall. With three phases of the cycle as against two, there are seven work roles as against three, and twice as many task groups. Yet overall numbers remain closely similar. The effect of changing from hewing to cutting is to subject the same number of people to a greater degree of social fragmentation.

The advancing phase of the cycle is unchanged, as are the roles of puller and stoneman, though their relations with other groups undergo considerable change (*vide* p. 58). The introduction of cutting handled in accordance with the principle 'one man—one task' has led not only to an additional phase of the cycle but to five additional work roles (six if the shotfirer is included), all narrowly defined and replacing the single role of hewer. Difficulties as regards the integration of task groups are likely, therefore, to be more acute on cutting faces, especially as isolate roles appear for the first time.

Yet, in view of the higher level of mechanization, one would expect a higher level of productivity. In the Area where the research was carried out it was the custom to fix cutting jibs at 4 ft 6 in. and hewing targets at 3 ft 0 in. per cycle. These norms, included in *Tables 1* and *2*, give a technical expectation that half as much coal again would come from the cutting as from the hewing face, for a just noticeable economy in manpower. No conventional cutting face visited showed a superiority of this order, no matter what allowances were made for existing circumstances, while much the highest production from a conventional longwall came from the hewing double unit described in Chapter XXIV. A close approximation to the full productive capacity of the cutting technology was made only under conditions of composite organization, on the faces to be described in Chapters VIII, XIII, and XIV, which reverse the conventional pattern. Though a conventional cutting longwall with a high performance record was visited during an earlier study, the results did not equal those of the composite faces and were obtained at the cost of a high level of tension, especially between the fillers and the various groups concerned with coal preparation.

Tables 1, 2, and *3* follow on pp. 50–52

TABLE 1 CONVENTIONAL HEWING LONGWALL ORGANIZATION
(North West Durham)

Average Unit Face Length 80 yds. Average Seam Height 2′ 0″.
Target Advance 3′ 0″ per cycle.
Total Cycle Group: Single Unit 23-25; Double Unit 40-44.

Cycle Phase	*Work Role*	*Type of Task Group*	*No. of Task Groups Single Unit*	*Double Unit*	*No. in Each Task Group*
Preparing and Getting (Two similar shifts)	Hewer	Identical	1, (2)	2, (1), (4)	12, (6), (24)[1]
Advancing (Third Shift)	Puller	Reciprocal	1	2	3, (4)[2]
	Stoneman	Reciprocal	3	4	2, 3, (4)[3]
2	3	2	5, (6)	8, (7), (10)	

[1] both shifts on same note represents standard; separate notes when shifts disagree; faces and shifts all together on double units in exceptional groups.
[2] an extra man according to custom and conditions.
[3] 2 in each tailgate, 3 in each mothergate caunch, sometimes 4 especially in fore caunch.

TABLE 2 CONVENTIONAL CUTTING LONGWALL ORGANIZATION

(North West Durham)

Average Unit Face Length 80 yds. Average Seam Height 2′ 0″.
Target Advance 4′ 6″ per cycle.
Total Cycle Group: Single Unit 22-24; Double Unit 38-41.

			No. of Task Groups		
Cycle Phase	*Work Role*	*Type of Task Group*	*Single Unit*	*Double Unit*	*No. in Each Task Group*
Preparing (First Shift)	Driller	Isolate	1	2	1
	Cutter	Reciprocal	1	2	2
	Scuffler	Isolate	1	2	1
	(Shotfirer)	(Isolate)	(1)	(2)	(1)[1]
Getting (Second Shift)	Hewer	Isolate	1	1	1
	Filler	Identical	1	2	7
Advancing (Third Shift)	Puller	Reciprocal	1	2	3, (4)[2]
	Stoneman	Reciprocal	3	4	2, 3, (4)[2]
3	7, (8)	3	9, (10)	14, (16)	

[1] ranks as official, not included in face team at workman level.
[2] cf. note [3] to *Table* 1.

TABLE 3 TRADITIONAL SINGLE PLACE ORGANIZATION

(North West Durham)

Average Face Length: 4 yds stall-side + 1 yd barrow way; or 6 yds rib-side.
Average Seam Height 2′ 0″.
Target Advance variable according to team.
Total Cycle Group: Bord Place 1-3: Winning Place 2-6.

			No. of Task Groups		
Cycle Phase	*Work Role*	*Type of Task Group*	*Bord Place*	*Winning Place*	*No. in Each Task Group*
Preparing, Getting, and Advancing (On all Shifts)	Composite	Identical[1] Reciprocal Isolate	1	1	3, (2), (1)[2] 6, (4), (2)
1	1	1	1	1	

[1] The task group has all three characteristics since, overall, team members do the same job, but reciprocate each other in picking up the cycle where it is left, and are on shift alone in bord places, though still members of a marrow group.
[2] numbers vary according to number of shifts worked in the pit.

CHAPTER VI

The Situation and Characteristics of Single Task Groups[1]

While status differences are acknowledged between faceworkers as a whole, underground datal men, and surface personnel, their existence among faceworkers themselves is apt to be denied. In single place working the members of a marrow group who shared all tasks had equal status but, with the advent of longwall specialization, status differences among the various task groups appeared, becoming more pronounced among the seven roles on cutting than the three on hewing faces. The purpose of the present account is to articulate the technical, sociological, and psychological distinctiveness of the various task groups. It will be served by selecting the main groups—cuttermen, fillers, pullers, and stonemen—concerned with the cutting longwall cycle as found in North West Durham. The functioning of the cycle group, which complements the present account, is described in Chapter XIII.

CUTTERMEN

A characteristic associated with low seam working that makes these longwalls different from conventional cutting faces seen elsewhere is the comparatively low status of cuttermen. Cuttermen in higher seams usually occupy an élite position,[2] which in North West Durham, both on cutting and hewing faces, is occupied by the pullers. With the mothergate breaking the face line, the cutter cannot cross from one face to another, so two machines are used, each face having its own permanent team.

[1] Source papers: 3, 5, 18.

[2] cf. Baldwin (1955) on cuttermen as a face aristocracy. He also mentions conveyor-shifters as having something of this status, but the puller role as known in Durham was not fully developed in the areas from which his main data were obtained.

With a distance of not more than 80–90 yds to undercut, the task becomes no more than a part-shift activity, the cuttermen spending as much as half their time on shift work. Under these conditions cutting is not an activity with which a man can completely identify himself. This has prevented its elaboration into a full specialist role.

Cuttermen, however, have compensations even with no more than a part-shift task. They have a large and powerful machine to handle, and their work is not confined to one place but takes them through the face. Like the pullers, they have the advantage of being a small group who complete an operation crucial for the cycle. They also work with each other, and with the scuffler, in a reciprocal way, all three remaining with the face for the duration of its life. Yet, though the two cuttermen choose each other and share a common note, the scuffler is paid on a note of his own and is separately cavilled to the face, so that an element of chance enters into the composition of the group which often gives rise to a lack of cohesion.

As the only face task calling for mechanical skill, cutting tends to attract men with some mechanical interest. Any qualified faceworker may compete for a vacancy, but the most usual entry is through scuffling, scufflers being recruited from any category of faceworker. Cuttermen must reconcile themselves to a working life that begins at either 3.30 a.m. or 8.50 p.m. (foreshift and nightshift times common in the pits studied). They are never on the backshift—the only 'good' shift.

The traits of men in cutting teams matched the situation described above, as though they had acquired the character of their role—displaying in the work situation certain attitudes which distinguished them from other task groups. This is a function both of the demands of the job and of personality. In the older mining communities particularly, men are well aware of the stresses each role imposes and of the opportunities it affords, and are free to compete for whichever they believe themselves to be most fitted. Cuttermen in North West Durham were able and intelligent but self-contained, quiet, and unassertive. From management's point of view they were never a source of trouble. Very different is the picture they present in pits where they compose a workman aristocracy, especially when one pair carries responsibility for the whole panel. Here, though their competence is granted, they tend to be perceived somewhat as 'prima donnas' and to some extent respond as such, surrounding their occupation with a mystique and making use of their power as a small, unique group to drive advantageous bargains as regards wages and privileges. Managers will say they must be able to depend on their cuttermen—and a high rate of reward in one special case does not unduly inflate total face costs.

FILLERS

Though the primary producers and the largest group, fillers have the lowest status. They are usually the entry grade to facework and, since other groups recruit from them, they contain a core of residual members. Like other main task groups, they are marrows but, unlike cuttermen and pullers who stay with a face throughout its life, they are cavilled every quarter. The quarterly cavilling of fillers, a feature which in some pits applies also to stonemen, is a carry-over from single place working. With increasing fractionation of the unitary collier role came permanancy of allocation to those occupying the new roles. The residual filling role was, however, still subject to customs and traditions, such as cavilling, applying to the earlier and more comprehensive role of collier. Hewers are similarly treated. The consequence is that the cycle work force is unstable, since the largest task group, the fillers (and sometimes also the stonemen), changes every three months. Fillers have a limited investment in any face to which they are cavilled and this is not conducive to the development of responsible relations with the cuttermen who precede them and the pullers who follow.

The low status of fillers derives also from the low degree to which the pullers and stonemen are dependent on their personal effort and capacity, for on cutting faces the amount of advance (which affects the pullers' and stonemen's piece-rate earnings) is determined by the length of the cutting jib. This may be contrasted with the higher status of hewers on hewing faces, where the advance achieved is not necessarily fixed by the cycle, but depends on their skill and effort.

Filling provides the least satisfying role on the face. The high level of interference arising from conveyor breakages and tub shortages is something the filler must learn to tolerate, in the same way as he must accustom himself to leaving his task uncompleted on many or even the majority of his shifts. Since he has less scope than those in other task groups for alternative work, equivalent either in cycle importance or financial reward to his own essential but repeatedly interrupted task, he must also accustom himself to inactive 'waiting-on'. A filler spends his time in a confined space shovelling coal, with occasional breaks to set supports. Unlike other faceworkers, he works on his own. Though he can hear other men along the face and see their lamps, he is without close neighbours. Added to his isolation is his feeling, arising from the interference to which he is so frequently subjected, that nobody gives particular thought to his troubles. On the faces visited, it was the habit of fillers to make a 'nicking' in the fired coal as one of the first activities of the shift, so that when their work was held up they had some shelter from the stream of cool air moving along the face. Each would retire into his hole and await events.

Their dominant attitude is a resigned acceptance of frustration. They complain, but believe there is little they can do. Having schooled themselves to accept dependence on circumstances and other people, they take little comfort when held up as the successors of the old colliers about whose feats of productivity stories are still told. The high rate of voluntary absence among fillers in the pits studied suggests that their work was scarcely attractive to them. The only attraction of the role is that it never involves night work, filling being always on fore- or backshift. Many young men prefer it for this reason.

TYPES OF FILLING TASK GROUP

Longwalls in pits visited elsewhere were often twice as long as those in North West Durham and had correspondingly larger groups of faceworkers. These panels were marked off in lengths or 'stints' which prescribed the amount to be filled off by each man. The most striking feature of the shorter Durham faces was the absence of stints, whose aim is to ensure that all members of a common payment group contribute equally when too numerous to remain in individual contact. *Group sectionalization*, an alternative way of dealing with this situation by dividing up the total face area into smaller payment units, though not widespread in Britain, is common in the coalfields of such countries as Holland where it is hierarchically organized, with a chargeman in control of each group. In Durham an equivalent was found with egalitarian features. On these double units, there were no more than 12-14 fillers, 6-7 on either side of the mothergate. These subgroups of 6-7 were each on their own paynote. In a group of this size the contribution of the individual can be directly assessed by his mates without recourse to formal partitioning. The same holds true in hewing groups. As such teams are self-selecting marrow sets, men of roughly equal capacity choose each other and mutual trust exists.

Under these conditions, there is an absence of the tension which characterizes large filling groups on the longer and more common type of stint face, where individuals of varying capacity are isolated in their lengths without the right to help from others. Informal partnerships between pairs or small groups too often prove unstable in these circumstances, while the *reactive individualism* which develops is a poor substitute for the support of a marrow group. This situation of unequal men in equal stints is inherently divisive when all are on the same paynote and the group is too large for one man's contribution to be known by the rest. Unequal stints and individual notes, which legitimize varying contributions and their corresponding rewards, have been accepted in some parts of the country and have led to improved results. Usually, however, they have been resisted, it being contended that the younger men who produce more should help the older who produce less and that all should share equally. Unequal stints can lead to unpleasant competition for places near the gates—which are more accessible to supplies—or for services from the deputy (cf. Baldwin, 1955 pp. 163-4); individual competitiveness is scarcely the most appropriate attitude to foster in a high risk situation. The Durham concept is that of equivalence of contribution as regards willingness and effort, rather than sameness of amount done, which is allowed to vary within the range accepted by the marrow group. At one pit, however, we found conditions where a demand for stints had been voiced. The men, cavilled to open up a new seam, had not previously worked together. There was much distrust and it was proposed to ensure equality of contribution by the imposition of stints. The idea was rejected by the lodge and not raised again once the men got to know each other.

In certain pits, filling is the last task of the cycle. In North West Durham, where pulling and stonework come last, should the filling get behind, coal left on can be taken off on the pulling shift. This possibility is ruled out when filling comes last. Such a difference in cycle position in conjunction with stints creates a stress pattern which may be summarized as *isolated dependence* in a *terminal position* (cf. Trist and Bamforth, 1951). As will be shown in Chapter XIII, however, the Durham filling pattern only partially mitigates the tendencies towards sub-optimum functioning arising from more general features of the conventional system. Nevertheless, among varieties of conventional filling organization, group sectionalization on an egalitarian basis makes the most constructive use of the marrow group tradition deriving from single place working.

PULLERS

On the North West Durham longwalls the pullers, because of the size of the faces, have combined advancing the face conveyor and gearheads with drawing off the supports, to establish a whole-shift role combining two of the five recognized face skills. These tasks require the reciprocal working of two, or sometimes three, men. Drawing off is generally recognized as the most dangerous facework activity, and within the small pulling task group there must be complete trust in each other's skill and judgement. This is reflected in the customs which regulate recruitment to pulling. Pullers are 'closed' groups, having complete control over the entry of new members. Others, who must already be experienced faceworkers, may become pullers only on invitation from an existing set with a vacancy.

On many shifts, the pullers come in to find the filling unfinished. Only they can 'save the cycle'. To do this they must complete the filling (at an agreed price) before going on to their own work. This means that the pulling role demands a wide range of face skills as well as willingness and ability to work hard and fast. These characteristics give it high status.

The behaviour required of men in the pulling role is very different from that characteristic of filling or cutting. To get through the extra work, they must be impatient of delay and not be put off by minor obstacles. They behave in this way. If a belt breaks while they are filling coal, they will mend it rapidly, and if their work is interfered with from outside the district, they will take positive action to find out the cause and get themselves going again. As would be expected, they take the same active and positive approach in other matters, and, because of the crucial nature of their contribution to the completion of the cycle, they are more successful than others in getting management to satisfy demands for special payment. As a consequence, the pullers, who are numerically a small group, stand out among the other task groups who have a more accepting and resigned attitude to their frustrations. Pullers keep, and indeed are kept by the others, rather apart; and this mutual aloofness is reinforced by the belief, particularly among fillers, that the substantially higher rewards of pullers are gained at their expense.

It was common in the Area in which the research pits were situated for the weekly earnings of pullers to be the highest among facework groups and those of fillers to be the lowest. The greater the puller-filler discrepancy, the greater the degree of cycle dysfunctioning, the stronger the bargaining power of pullers and the weaker that of fillers. Though the actual earnings of pullers tended to be a rather closely guarded secret, a sense of relative deprivation[1] would sooner or later drive the fillers to take steps to improve their own rates. This in turn would induce similar movements in the intervening groups. The differences in status, power, and reward between the various single task groups, exaggerated by cycle dysfunction, generated continuous pressures for piecemeal wage gains. This aspect is further discussed in the next chapter.

STONEMEN

Stonework is done in the gates which, apart from the fore caunch, are work areas not shared by other task groups. When there is more than one group of the same kind, as with pulling or

[1] For this concept *vide* Stouffer *et al*, 1949.

cutting, these are of equal importance to their own particular phase of the cycle. Stonework groups on the other hand form a hierarchy related to the differing dependence of the cycle on the different caunches. Fore caunch men have the highest status; the face cannot advance if they do not finish. Back caunch men come next; stones left in the gate need not hold up progress though they cause inconvenience. Tailgate men come last; theirs is work which can be left for a whole shift if the occasion arises. These differences are sometimes, though not always, reflected in the prices offered for work in the various caunches, especially where these are 'let' as separate 'bargains'. The separation of the caunches in this way restricts mobility and prevents the full complement of stonemen from becoming a cohesive group.

Through the tendering system, the four caunches are often let as bargains, several groups of stonemen putting in tenders and the choice being made by the undermanager. The price tendered and accepted for each caunch is in nearly all cases a standard 'seam price'. In theory, these bargains are re-tendered every quarter, when a change can be made, but groups usually stay with a caunch for the whole life of a face. When caunches were let as bargains, the status of stonemen was between that of pullers and cuttermen. By contrast, on a hewing face at another pit, stonemen had the lowest status, though the three types of caunch were in the same order. This was because the composition of gate teams depended on chance, men being cavilled to caunches from those stonemen in the seam—including backbye—who were eligible for the next set of face places. Such a practice gave stonemen only temporary facework status.

The situation of stonemen is more like that of a single place group than of a task group in the longwall system. Being relatively independent of the sequence of operations in the face itself, they have greater scope for uninterrupted task completion and correspondingly greater work satisfaction. They regard the caunches for which they have tendered as their own habitats. The physically demanding nature of the work brings them high respect. Of all task groups they were the most autonomous and the least complaining. Though relatively isolated from other faceworkers, they were not aloof like the pullers, or dispirited like so many of the fillers, but cheerful and outgoing in their relations.

The main drawback is permanent nightshift. This stonemen do complain about, though they tend, publicly, to shrug it off as an unavoidable disadvantage that goes with their chosen occupation. Nevertheless, continual night work produces irritation and frustration in other areas of their lives, of which their high voluntary absence is a sign.

CHAPTER VII

The Displacement of Operational Control[1]

THE PROBLEM OF CYCLE CONTROL

Illustrative of the distinctiveness of its constituent task groups and its general structural complexity is the example set out in *Table 4* of the cutting longwall cycle aggregate, which is further discussed in Chapter XIII. The principal difficulty in operating such a system stems from the need to integrate the miscellaneous collection of segregated task groups into an overall organization for the performance of a cycle, which, technologically, is an interdependent whole. The more each task group is built up round its own paynote, the more it tends to acquire a goal of its own. Separated still further from other groups by differences in psychological climate, social character, internal structure, and method of recruitment, each relates the cycle to itself rather than itself to the cycle. Longwall task groups have been placed by the structure of the conventional system in a situation which invites them to be primarily concerned with improving their own position.

Since the various groups do not operate in a system where they can cohere spontaneously, co-ordination has to be externally supplied by management. The whole onus of cycle control, therefore, falls on supervisory authorities outside the face group. Yet these have remained the same in title, and very much the same also in the definition of their functions, as their counterparts in single place working where the system was largely self-regulating. A deputy has shift responsibility for a face district; an overman co-ordinates all activities in a seam, also only during one shift; and an undermanager effects unified control over the whole cycle but at three levels from the coal face. Officials feel, in a way they find both hard to pin-point and hard to endure, that they get no

[1] Source papers: 5, 20.

Two 80-yd faces, one each side of mothergate

Shift Sequence	*Work Location*	*Task Group*	*No. of Men*	*Method of Payment*	*Type of Role*	*Allocation to Face*	*Recruitment Route*	*Kind of Interference*
Fore or Late Night	Face A	Cutting Scuffling	2 1	Per Yard ,,	Reciprocal Isolate	Permanent ,,	Scuffling Direct	Geological conditions, power and mechanical breakdowns
	Face B	Cutting Scuffling	2 1	,, ,,	Reciprocal Isolate	,, ,,	Scuffling Direct	
Back or Fore	Face A Mothergate Face B Panel	Filling Hewing Filling Drilling	6 2 6 1	Tonnage ,, ,, Per Hole	Identical ,, ,, Isolate	Qr. Cavil ,, ,, Permanent	Hewing Direct Hewing Filling	Tub shortages, conveyor breakdown, geological conditions
Early Night or Back . . . . Late Night or Early Night	Face A Face B Tailgate A Fore caunch Back caunch Tailgate B	Pulling ,, Stonework ,, ,, ,,	3 3 2 4 3 2	Per Operation ,, Cubic Measure ,, ,, ,,	Reciprocal ,, ,, ,, ,, ,,	Permanent ,, Qr. Bargain ,, ,, ,,	Invitation ,, Direct ,, ,, ,,	Geological conditions
First Shift	2 Locations	2 Roles 4 Groups	6	1 Method	2 Types	1 Method	2 Routes	*Notes* Shift sequence alternates weekly. Driller works a split shift, overlapping filling and pulling. Pullers start earlier than stonemen — hence early and late night shifts.
Second Shift	1 Location (Panel-Wide)	3 Roles 4 Groups	15	2 Methods	2 Types	2 Methods	3 Routes	
Third Shift	6 Locations	2 Roles 6 Groups	17	2 Methods	1 Type	2 Methods	2 Routes	
3 Shifts	8 Locations	7 Roles 14 Groups	38	5 Methods	3 Types	3 Methods	5 Routes	

support from the men in their effort to maintain the cycle. The feelings of disappointment, rejection, and anger that this experience generates discolour relations. They constitute the officials' grievance as distinct from the miners'. On the present analysis this grievance has no effective remedy within the conventional longwall.

THE USE OF THE WAGES SYSTEM

Given the structural impossibility of self-co-ordination on the part of the task groups, there were two essential methods by which management could seek to control the cycle: direct coercion and price negotiation. The nature of work in the underground situation is such that a close regimental type of supervision has never been feasible, apart altogether from the resistance it would arouse. Coercive control has only limited usefulness. The method of price negotiation was established as a principal feature of the single place tradition (Chapter IV) and an attempt has been made to handle the very different longwall situation predominantly through the same control mechanism. This, however, has led to such an expansion of the bargaining process, especially about sub- and ancillary tasks—which have to be done if the cycle is to be completed on schedule—as to become dysfunctional.

Management through the wages system—the use of a system of rewards for operational control—has developed to a far greater extent in mining than in most industries (though a comparison with certain types of fabrication beset with detailed rate fixing difficulties would be illuminating as regards both the similarities and the differences). The general result is that every detail of the production process has become affected by the balance of power between management and workers. Unless this is understood as a system characteristic of the conventional longwall, labour relations in the industry, especially since nationalization, must remain enigmatic.

In this research our concern is not with wages in themselves but with the way in which wage practices reflect the pattern of longwall organization. It is the above aspect, therefore, which has been isolated. As in single places, the basic concept is the price agreement—a price formally negotiated between management and men for a specified amount of work in a task of defined scope—within a range of conditions accepted as applying. As soon as these boundaries are transcended the question of a new price arises and fresh negotiations begin. In single place systems, however, there was comparatively little task breakdown and price agreements remained reasonably holistic in relation to comprehensive work roles in an environment where there was little cycle disturbance and a perceivable relation between effort and reward. With the greater differentiation and degree of disturbance brought in by the longwall, separate agreements have been negotiated for an increasingly long list of activities so that *itemized price lists* have come into existence. Not only main tasks but sub-tasks and ancillary activities are subject to separate agreements. Though the prices themselves are usually settled in negotiation, the amount actually done by a particular work group is often impossible to determine with exactitude, and direct bargaining takes place about how much should be paid, usually underground between men and an official with notebook and pencil, hence the term to be 'paid on the pencil'—on a system based on group pressures rather than task completion.

When a system of wages is based on detailed job breakdown it becomes impossible to find a common denominator for fixing rates. In the present instance several different criteria are used—tonnage, yardage, cubic measure, number of operations completed, etc. As a result, contradictory interests develop. When one aspect is selected for payment, the task as a whole becomes divided into rewarded and unrewarded parts and the latter invite neglect. For such neglect it is impracticable to impose penalties except in the grossest cases. Every task group can, in a number of ways, seek special payment for additional labour entailed in putting right what has been neglected or left incomplete by other groups. In this way vicious circles are created. The bargaining, minor disputes, and claims entailed and the tension engendered deflect both men and officials from the primary task. The time and energy consumed can be enormous, preoccupation with the process becoming a chronic substitutive activity which, as it continues, becomes accepted as 'normal' (*vide Table 13*, p. 127).

THE END RESULT

Management on the face itself is still represented by the deputy, whose role derives historically from statutory requirements concerning safety and whose traditional relationship with the face-work group has been one of service rather than of operational leadership. Attempts to transform him into an operational leader have on the whole remained aspirational, and the most important recent development has been to require him, rather than the overman, to pay the men. Though this allows more to be settled on the spot, it also exposes the most junior level of supervision to immediate group pressures. This step, nevertheless, represents the logical end-state of management through the wages system and gives the pattern both completeness and self-consistency in the longwall situation.

Management roles have remained embedded in the structure developed for single place working. The principal development has been the elaboration of the price list and the bringing of its administration nearer the coal face. Yet the scale of forward planning, the standard of maintenance, and the accuracy and speed of 'intelligence' necessary to achieve the co-ordination required to give longwall faces effective service and to keep them free from internal and external interference are of a qualitatively different order from those required by earlier systems. As in single places, there is no unification of control below undermanager, deputies, even when they overlap, having only shift responsibilities, as do overmen. Yet with a formal cycle now spread over three shifts, such a situation inevitably breeds disorganization unless the cycle group, through the pattern of its own internal organization, is capable of exercising a high degree of self-regulation.

The general situation may be summed up as follows. Longwall systems, because of their greater degree of differentiation, require much more integration than single place systems; but the conventional pattern of organization has broken up the traditional, self-regulating cycle group into a number of segregated single task groups each bound within its own concerns. These groups depend entirely on external control in order to carry out the indivisible primary task of completing the cycle. The existing pattern of management through the wages system can only partially supply this control. Full control would require either a degree of coercion which would be both impracticable and unacceptable or a degree of self-regulation which implies a different organizational pattern.

SECTION THREE

Emerging Forms of Work Group Organization

CHAPTERS

CHAPTER VIII

The Emergence of Composite Longwall Working[1]

THE COMPOSITE TRADITION

In view of the difficulties of integration encountered on conventional faces, great interest attaches to longwall systems which have developed a form of work organization deriving from, rather than running counter to, the single place tradition. In these systems, which are known as *composite*, a common task and a common paynote reunify the cycle group.

Before the present research, the Tavistock Institute had encountered only one basic form among several varieties of composite working, in which all activities were carried out in one shift. Because this limited face length, it was said that these methods had restricted application. Three coal-getting shifts could be run in the 24 hours, though in most pits the number was reduced to two in view of winding problems and on-cost personnel. One such innovation emerged in East Midlands Division as part of a plan to introduce continuous mining (Sheppard, 1951). While a new cutter-loader was under development a composite system came into existence in which handfilling was retained. Faces were concentrated groups of single units, averaging 100 yds in length, with adjacent panels won on advance and retreat. Faceworkers were multi-skilled, the cycle group of 25-30 being on a common note and three such groups sharing a face. There was at first a day wage, but when the system had to take account not only of its own fully mechanized form but of cycle systems using power-loaders, a group bonus was added.

At two of the single place pits visited in North West Durham

[1] Source papers: 8, 21, 23, 24.

there were cutting faces, 30–60 yds in length, which were manned by one small group who shared a common note and carried out all operations. These faces were called *composite shortwalls*, and represented the application of the single place tradition to short conveyor faces. A group of six to eight men worked a face for one shift during the 24 hours, taking up the cycle where they had left it the previous day. Being multi-skilled, they deployed themselves as required to cutting, filling, drilling, pulling, and stonework. Since these faces were located in seams where single place working was carried out, the cycle was not constrained by the need for completion at any particular time.

At yet a third pit, cutting shortwalls were visited where fifteen men, sharing a common note, worked each face for three shifts—five men per shift. These shortwalls were enlarged rib and stall workings, in which a 30-yd face was advanced to a boundary and a similar 30-yd face retreated. The East Midlands units represented a still further extension of the same principle. All operations were generally completed in one shift and the next cycle was begun by the same men. When the face was retreating and no stonework was required two cycles were often achieved. The men were experienced workers who had marrowed themselves into sets of fifteen. Each group developed its own method for allocating its members into sets of five and for rotating shift times every week. In this way they avoided the longwall situation in which a number of men were tied to one main task carried out only on 'bad' shifts. The basis of payment was the same as on conventional longwalls, each operation having its own piece-rate. The total amount so earned was credited to each fifteen-man team and then divided equally among all members.

A similar system, with 40-yd retreating faces and seven to eight men on at any one time, had been introduced in a drift belonging to the Yorkshire pit where the conventional longwall had first been studied. These faces gave the senior author and his then colleague, K. W. Bamforth, their first experience of composite working and a brief chance to observe the effects on men changing over from conventional methods. A new group cohesion was experienced; men looked less fatigued and felt under less strain. Older workers said they could have a longer working life at the face and compared their experience of the new system with that of bord and pillar working or step-wise longwall. Composite shortwalls embody three features characteristic of single place systems—task versatility in faceworkers, a sense of mutual responsibility between members of different shift groups, and shift rotation on an egalitarian basis. Several workers said they were recapturing qualities of relationship and satisfaction which they had lost since working in earlier systems. There was also less trouble with management, and production was maintained at a high level. In East Midlands, where similar attitudes were reported from a number of face groups in six different pits, the relevance of composite organization to higher mechanization became apparent. But an opportunity to compare orthodox with composite work organization on otherwise identical longwall faces arose only in the present study, as did that of tracing, through living examples, the derivation of the composite from the single place tradition.

THE MANLEY INNOVATION

In one of the Durham collieries shortwalls had been introduced when roof conditions in the Manley seam made long faces impossible to support. Unfortunately, the price agreement was unrealistic and men had to be 'made-up' to an adequate wage. Eventually the increasing cost of datal labour needed for winding on all three shifts and the amount of unproductive stonework created such dis-economies that management proposed a return to conventional longwall. Among the men the belief persisted that the seam could not be safely worked with long faces. Moreover, they did not want to go back to a system which tied them to particular shifts and single tasks. The problem of roof control was solved by an improved chock, but it took a year of negotiation between management and lodge before an agreement was reached on how to preserve the socio-psychological advantages of composite work groups and yet solve the economic problems which necessitated longer faces and the three-shift cycle.

On this background a major development took place in the application of the composite approach to the orthodox longwall. It resulted, under the conditions prevailing, in self-selected cycle groups of 41 men, who allocated themselves to tasks and shifts and received a comprehensive payment on a common note. The tasks and shifts represented the orthodox longwall sequence and the numbers required for each activity were agreed in detail between management and lodge. The wages side of the agreement was in two parts. The first, which accounted for 58 per cent of the note, was a basic wage set above the legal minimum with the intention of removing all need for making-up; it included a comprehensive payment for sub-and ancillary tasks so that bargaining with the deputy was virtually eliminated. The second part was a 42 per cent bonus on output, which was earned in full if, without reinforcement, the cycle was completed.

In common with other arrangements of the same type, the pit's formal agreement for composite longwall conveyor working prescribed the technical conditions for completing the cycle and the basis and method of payment. No reference was made to work organization, but the inclusion of the word 'composite' was meant to embody established customs. The continuance of these was implicitly assumed and was ensured by the basis and method of payment and by the experience and attitude of the men. As regards rotating shifts and tasks, the way in which facework groups would organize themselves was subject to an understanding arrived at between management and lodge during the negotiations, though this was not embodied in the formal agreement (*vide* Chapter XV).

The conception of retaining the three-shift cycle, while proceeding in accordance with composite principles, was novel. The research team had been informed by people at widely different levels in the industry that it could not be done. Nevertheless, an approximation had been described by Baldwin (1955, p. 167) as widespread in Warwickshire, which brought fillers together with packers and drawers-off under one 'all-in' contract with interchange of shift and task. Similar arrangements were mentioned by both men and officials as occurring in parts of Yorkshire. In all these cases, however, cuttermen, pullers, and stonemen were left out. More task groups are included in longwall composite teams at a somewhat higher level of mechanization in the Donbass (Report by the Technical Mission of the National Coal Board, 2, 1957). Yet even here 'caving' teams, repair men, and gateway stonemen are omitted, with disadvantageous consequences according to those concerned. If the total three-shift group had indeed been elsewhere included, no other pit at the time of this study appeared to have worked out a satisfactory agreement. Several informants suggested that 'something like this' had been tried in other pits only to fail because of difficulties over establishing a workable agreement. They were not, however, able to give chapter and verse. No other successful cases were located in the Division and the research team was soon to have direct experience of the circumstances leading to failure—which were distinct from the method itself and the principles on which it is based.

During the research at the colliery, composite agreements were drawn up for each of four seams. Though the formal agreements differed in minor detail only, work organization showed wider differences, partly for geological and technological, but mainly for socio-psychological, reasons. A detailed study was made of two double-unit cutting longwalls comprising the major part of one seam and of a hewing double-unit comprising another. A less detailed study focused on the changeover to composite working in yet a third seam. The next chapter, which gives a qualitative account of the general characteristics of composite systems, is concerned with cutting longwalls, not only because most of our work was carried out on such faces, but because they represent the most differentiated type of longwall technology and illustrate the problems of cycle co-ordination and task group cohesion better than do the less differentiated hewing longwalls. Chapters XIII and XIV evaluate examples of composite systems in operation, and subsequent chapters describe the evolution of their characteristic forms of work organization.

CHAPTER IX

The Nature of Composite Self-Regulation[1]

THE BASIC POSTULATES

Composite longwall working depends on successfully binding together into a corporate whole the rather large overall group comprised by the successive task groups of the three-shift cycle. Widespread scepticism exists that this can be effectively done. It is argued that in the conventional system most trouble occurs in large groups of hewers and fillers, and that among the smaller marrow groups performance varies widely, so that, if merged together, either the pace would be that of the slowest or the 'merger' would split up through internal disagreements. There is force to these contentions, but they overlook a fundamental point: the members of filling and hewing groups perform identical as distinct from reciprocal roles. The same is true of the marrows who succeed each other in single place working, though in their case the task is holistic rather than fractionated. In all these groups problems centre on the different amounts of the same task alleged to be done by various members. Problems of composite longwall organization concern the relating together of individuals performing different tasks—the interdependent phases of the cycle in which roles are reciprocal. The size of identical task groups is not increased by the application of composite principles to the orthodox longwall. The overall cycle group which comes into being is composed of the same activity groups which previously had only separate existence. Under these conditions the basic assumption of the wages system becomes organizational as distinct from simply quantitative as in conventional longwalls—where common notes are shared by those who do the same tasks—and the payment group becomes a multiple as distinct from a single role group.

[1] Source papers: 7, 11, 12, 13.

Such a group is willing to share a common note so far as all members believe that the level of their personal earnings is kept up by regular completion of the overall task—perceived as an outcome of interdependence rather than of separate achievement. To pool total earnings, however, does not necessarily mean that equal rates must be paid; this follows only if all contributions are regarded as equal, which entails the further assumption that optimum cycle functioning depends on interchangeability of workmates as well as on interdependence of work roles.

The 'basic postulates'[1] made by the groups on the Manley panels conformed to this pattern which derives directly from the single place tradition. The essential concept is that of the multi-skilled faceworker interchangeable with others according to need, all such men being regarded as making an equivalent contribution and therefore as meriting the same reward. The common fund of underground skill and the common identity of being a faceworker have been made primary. The different skills appertaining to the different tasks of the production cycle and the separate identities that build up around them have become secondary. The emphasis of the conventional system has been completely reversed (cf. *Figure 5*).

FOUR INTERRELATED ASPECTS

Four distinct though interdependent aspects of composite longwalls may be identified: the method of work, the kind of workman, the kind of work group, and the form of payment. Each of these is described below as found on the Manley panels. They compose the common theme on which the differences that developed are variations.

Composite work method. As in single place working, oncoming men take up the cycle at the point left by the previous shift. When the main task of their own shift is completed they redeploy to carry on with the next, whether this forms part of the current cycle or begins another. This is the central feature of the composite work method, which will be referred to as *task continuity*. Preparedness to carry on spontaneously with the next task of the cycle distinguishes composite from conventional

[1] For this term *vide* Hoebel, 1954.

working. More than on anything else its effectiveness depends on the extent to which the next task is continued as soon as the main task of a shift is finished. Such continuity of activity prevents lag from developing in the earlier phases of the cycle and provides 'internal reinforcement' should difficulties arise later. There is rarely any need for external reinforcement and the cycle group becomes self-maintaining.

Composite workmen. How far task continuity can be practised depends on whether the cycle group includes men competent, if not always formally qualified, to undertake, under supervision, the necessary tasks as they arise. Composite workmen are multi-skilled miners qualified at least in filling and stonework, and often also in pulling, drilling, and cutting. Though being a member of a composite team does not necessarily extend the range of a man's skill, it tends to maintain his experience. How far it does so depends on which main tasks he carries out, on which parts of the face he works, and on the frequency with which states of cycle lag or advance arise when he is on shift. Though all members of a cycle group do not need to be competent in all five of the recognized face skills, the team as a whole must be able to deploy sufficient resources on each shift to man the roles likely to arise.

Composite work groups. The team manning a composite longwall is self-selected so that the range of personal abilities is accepted from the start. In this respect it has affinity with the marrow groups of single place systems and the single task groups of the conventional longwall. In composite systems the marrow relationship holds throughout the entire cycle group. Such an autonomous group accepts responsibility for allocating its members to all roles that management requires them to fill in order to complete the prescribed task. To regulate deployment, the team works out its own system for rotating tasks and shifts. This ensures to individuals equal opportunity of good and bad work times and demands in return reasonable equivalence of contribution.

Composite payment. As in single place systems, there is a common paynote in which all members share equally, since all are regarded as making an equivalent contribution. The note is based on an inclusive shift-rate covering all work done at the face. The items paid for separately on conventional faces, which give rise to so much bargaining and argument, are consolidated into a comprehensive amount which is higher than the minimum. An incentive bonus is then added based on coal produced. The only additional items arise from causes outside the control of the face team—money for waiting-on, wet money, etc. This system of payment places on the team as a whole responsibility for all operations at the coal face, while relating total earnings to productivity and removing the need for making-up by guaranteeing a basic wage.

THE KEY FUNCTION OF TASK CONTINUITY

For a system to become 'completely composite', all four aspects require full development. Nevertheless, only one—task continuity—is *sine qua non* in all circumstances. Provided task continuity is consistently practised, the other three components may take various forms or be present in lesser degree. The basis of payment may be a price for cycle completion, or a day wage, as found at first in East Midlands. Unequal division of the common note may, under certain conditions, be more appropriate and was practised on the Donbass faces referred to in the Report of the Technical Mission and on the Dutch faces described in Chapter X. It is not essential that as many men should be multi-skilled as on the Manley panels or that so elaborate a system of task and shift rotation should develop—as was discovered later in Durham—though, of course, these features are highly advantageous.

By contrast, the regular practice of task continuity is fundamental for maintaining cycle progress. *Table 5* describes in greater detail than previously given the component activities of the cycle, the relations between which are shown dynamically on *Figure 4* as a succession of nodal points of divergent and convergent dependence. Given such components and the structure of their technological dependence, *Table 6* summarizes the progress to be expected at the end of each shift under conventional and composite organization. The norms described were tested by repeated observation on the faces discussed in Chapter XIII and checked by

TABLE 5 CUTTING LONGWALL COMPONENT ACTIVITIES

Main Task	*Component Activities*	*Location of Activity*
CUTTING	(a) *Turning and changing picks*, replacing worn picks on the cutter	Face
	(b) *Jibbing in* the cutter	,,
	(c) *Undercutting* the coal, timbering the face as cutter proceeds, setting nogs, scuffling	,,
	(d) *Turning the cutter* at the end of its run	,,
	(e) *Shotfiring* the undercut coal (by an official)	,,
FILLING	(f) *Tracking through* the shot coal	Face
	(g) *Breaking in* to the face	,,
	(h) *Filling off*, face timbering, clearing up spillage, and disposing of band	,,
	(i) *Hewing* in the mothergate (also during cutting)	Mothergate
	(j) *Coal drilling* commences during filling, continues with pulling and stonework	Face
PULLING	(k) *Breaking and pulling* the belt conveyor and tension end	Face
	(l) *Pulling up* the conveyor gearhead	Mothergate
	(m) *Coupling up* the belt, replacing bars, tensioning the belt, setting the face signal	Face
	(n) *Drawing off* the waste, resetting chocks and props	,,
	(o) *Extending* the mothergate belt on alternate days	Mothergate
STONEWORK	(p) *Stone drilling* (commences after (l)), placing extra supports, erecting fore-poling, firing the shots (by the deputy)	Mother- and Tailgates
	(q) *Pack building* in the gates and stowing in the goaf	,,

TABLE 6 CYCLE PROGRESS AND TASK CONTINUITY

Shift	*Conventional*	*Composite*
I	*Activities (a)–(e)* When cutting is finished, personnel are sent on to shift work. Segregated work groups and divided notes make it impossible for cuttermen to touch the fillers' coal.	*Activities (a)–(g)* After cutting is finished the coal is fired while the men have their bait. There is no barrier to their going on with tasks preparatory to filling, which constitute the second phase of the shift.
II	*Activities (f)–(j)* All preparatory tasks for filling have to be done as well as filling itself. If interruptions or breakdowns lead to much waiting-on, coal is left on the face and cycle lag develops.	*Activities (h)–(m)* Filling can start immediately and is usually finished in time for all work preparatory to pulling to to be done during a second phase. Considerable delays can be borne without cycle lag developing.
III	*Activities (k)–(q)* At best, seven activities remain. If coal is left on, the pullers get this off as an extra, making additional money which gives them a vested interest in cycle dysfunction. Reinforcements have often to be drafted in to help the pullers complete the filling, and also at times the pulling. If a lot of coal is left on, the cycle may be abandoned; the oncoming men are then put on shift work as are the cuttermen following. If caunch work gets behind, it must wait till the next stonework shift.	*Activities (n)–(q)* No more than four activities remain. Rarely is any coal left on, and the pullers, having usually had their preparatory tasks done, can provide internal reinforcement to the stonemen during a second phase of the shift. There is, therefore, no need for external reinforcement to complete the packing. Such tasks as extending the mothergate belt are incorporated into the work of the face team. Should caunch work get behind, it can be picked up at a convenient point during the next cycle.

FIGURE 4

DEPENDENCE STRUCTURE OF TASKS ON DOUBLE-UNIT LONGWALL (simplified)

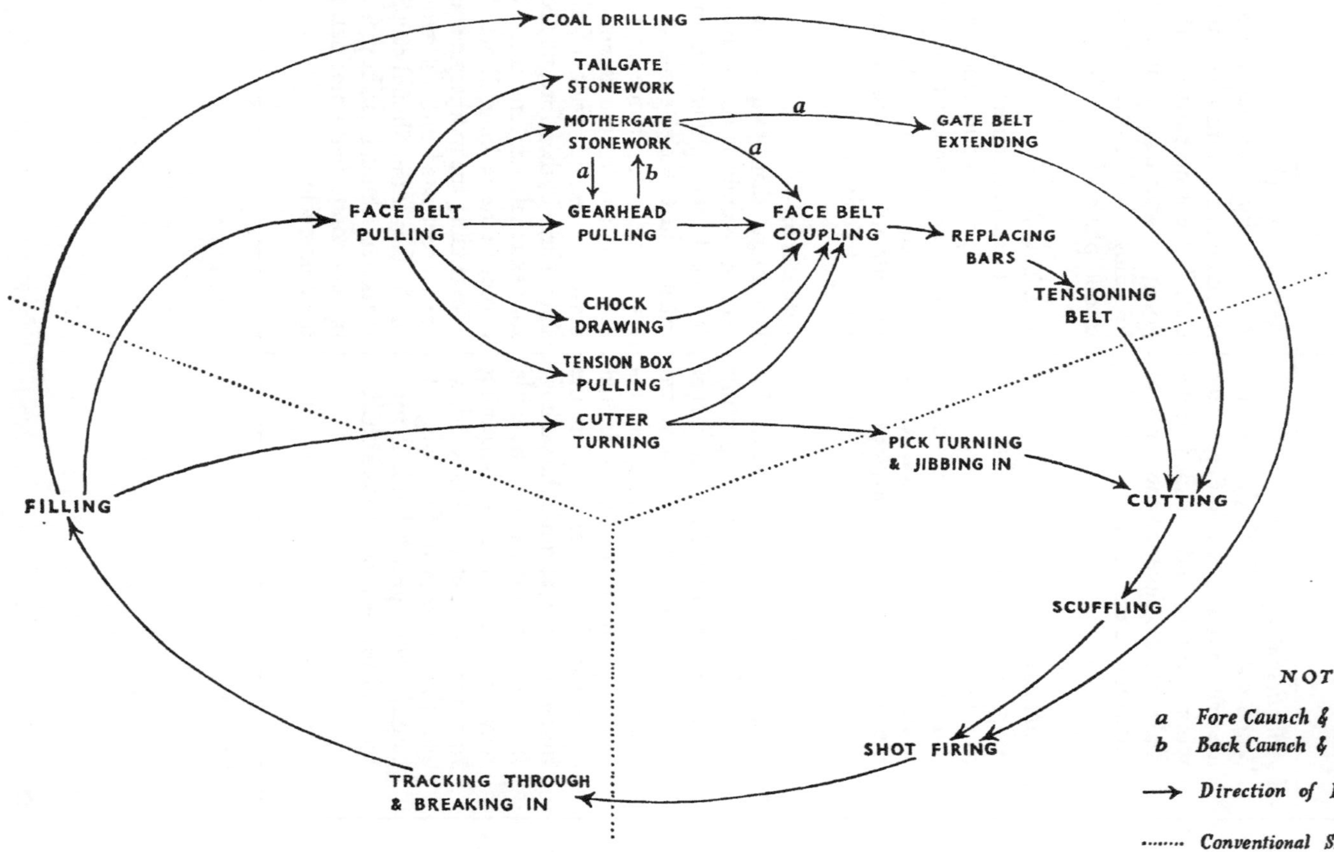

NOTES

a *Fore Caunch & Leading Face Gearhead*

b *Back Caunch & Lagging Face Gearhead*

→ *Direction of Dependence*

······· *Conventional Shift Boundaries*

FIGURE 5

TECHNICAL AND SOCIAL RELATIONSHIPS WITHIN CUTTING LONGWALL TEAMS WITH (A) CONVENTIONAL AND (B) COMPOSITE TYPES OF FACE ORGANIZATION

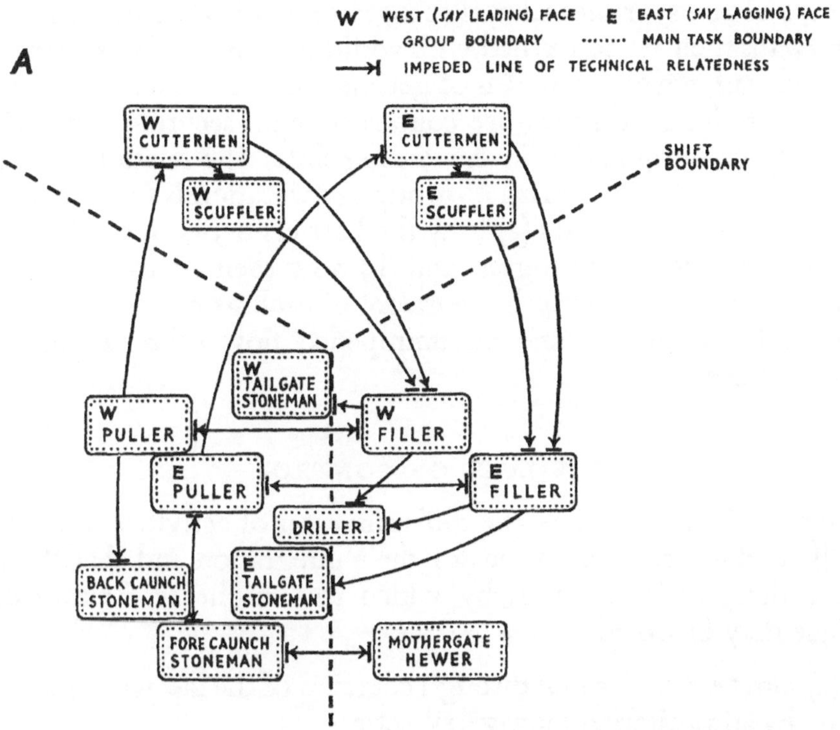

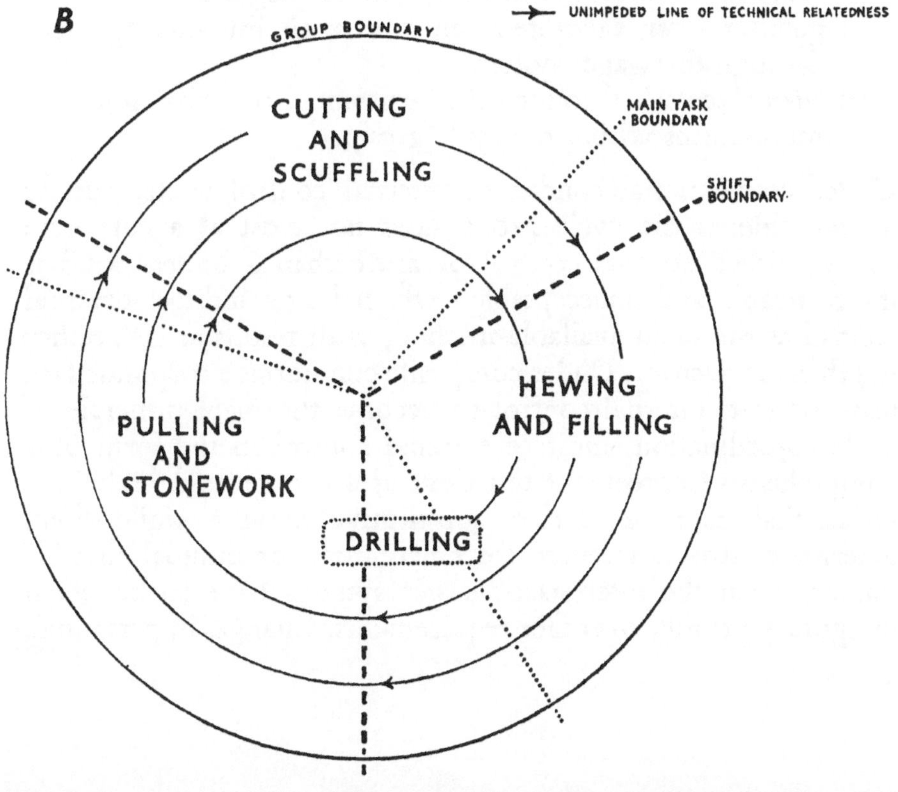

the use of an experimental cycle progress record. On conventional longwalls men are concerned with additional tasks only as a result of cycle lag, never as a means of getting the cycle in advance. On composite longwalls they are concerned with securing just such an advance—originating in the cutting shift. The relation of this difference to the two forms of group organization is presented in *Figure 5*, which indicates (a) how the barriers between single task groups impede cycle progress and (b) how their removal pushes the sequence ahead with the 'horizon' of each group extended to the cycle 'perimeter' and an unimpeded flow of relationships taking place.

THE LOCI OF CONTROL

Chapter VI showed how the differentiation of activities in longwall working created a need for their integration and that there were three main methods by which this might be attempted. These may be called

(a) *direct external control* through coercion of the facework group by authoritarian managerial action
(*b*) *indirect external control* through detailed negotiation and bargaining over each item on the price list—management through the wages system
(*c*) *internal control* through the existence of self-regulatory mechanisms within the cycle group.

Under conventional conditions internal control is structurally unrealizable as the cycle group does not exist as a corporate entity; direct external control of more than a limited kind is impracticable and unacceptable—which leaves indirect external control as the main available method, with results at best rather less than satisfactory. Under composite conditions a corporate face team permits internal control to become the main principle of cycle co-ordination. Indirect external control in the form of a comprehensive agreement functions as a background 'stabilizer' rather than as a 'breeder' of substitutive activity, while direct external control is reserved for emergencies or unusual circumstances when the interests of larger systems have to be given temporary priority over face requirements. Changes in personnel, equipment, or the task environment which produce disturbances of a more than temporary character lead to a re-appraisal of the framework of indirect external control at colliery level—by management and lodge. In this way modifications are sanctioned which enable the face system once again to become self-regulating.

The relations between direct and indirect external control and the internal control exercised by its members express the relations of the face team to the larger systems in which it is included and to the resources of its own component activity groups. The more capable the latter, the more complete can self-regulation become, while their weakness necessitates more action from outside systems. Sections Five and Six present contrast studies in the balance of these relations.

IDENTITY OF AIMS

Since men and officials on composite longwalls share the common goal of cycle completion, they can co-operate in a way not possible when task and cycle responsibilities are split from each other in terms of what 'belongs' to management and what to worker. Identity of operational aim removes the 'officials' grievance'. With the support of a self-regulating cycle group, the deputy has more scope for using his knowledge and experience of mining than when he is entrammelled in the wages system, and on composite faces the idea of his becoming an operational leader is no longer unrealistic. When men are paid only for the amounts they do of selected aspects of single tasks, they take any attempt to advise as implying that they are not up to their job and therefore not worth their money. Similarly, any request to do anything additional is regarded as exploitation unless separately rewarded. This situation deprives face officials of any reasonable chance of exercising technical leadership and leads to the perpetuation and concealment of inferior work practices and to resistance in learning more apposite skills. A comprehensive agreement which commits a corporate group to an overall task legitimates motivation to improve performance and releases ability to learn. The technical advice of the deputy is sought, and the leads he offers are willingly tried out. His mining judgement is welcomed not merely from the point of view of maintaining a threshold of safety but from that of making the most of existing conditions or dealing in an effective way with change. Similar demands are made on senior officials, who have the opportunity of making constructive managerial use of pressure towards greater technical effectiveness arising from the primary group itself.

Technical leadership increases in importance as mechanization proceeds. Its enhanced scope, however, was already apparent on the semi-mechanized composite faces first visited by the research team during the preliminary studies, where the improved quality of relations with officials was in striking contrast to the atmosphere prevailing in conventional settings and had come about for the reasons described. A task-oriented group seeks technical leadership which furthers its mission, while an itemized price list negates mission commitment, and coercion denies that management and the working group can have a common aim.[1]

THE DISTRIBUTION OF LEADERSHIP

In exercising internal control over the cycle, the Manley groups did not make use of specific men as group leaders to more than a minimum extent, though under conditions of higher mechanization the institution of formalized internal leadership may be required (cf. Chapter XI). Team captains were representatives rather than executives, and the group spread cycle responsibility equally among all members. Every man carried out whatever he was doing in such a way as to further the work of others. In the conventional system, the limit of his responsibility being the boundary of his own task, he is not constrained to do this. In the composite system, this limit being the cycle as a whole (*Figure 5*), he is induced by the nature of the 'force field' (Lewin, 1951) to work in a way that furthers the general objective.

Their predominantly multi-skilled qualifications permitted the Manley workers to develop systematic methods for the rotation of shifts and tasks (Section Five). Not only did this allow them to maintain their skills, it gave a continual reminder of the conditions under which each main task had to be done and of the consequences of sub-standard work. The rotation of shifts and tasks developed a common standard of workmanship, which in turn

[1] cf. Gouldner (1954).

acted as a mechanism of cycle control. Having experience of all phases, men were more readily able to evaluate the state of the cycle when they came on shift and could more easily adjust their work pace and the pattern of their deployment to the requirements of the situation.

While the practice of task continuity is the minimum requirement that must be met before a composite system can develop, the range of variations beyond this minimum is considerable, in quality as well as quantity. There are types and degrees of compositeness; for the best outcome, type must be appropriate to technology and degree complete—in all dimensions. In the case studies presented in the later parts of this book comparisons will be made of the main variations found during the research, together with an attempt to determine their position on a 'scale of compositeness'. Attention will also be paid to the forms most suitable for higher levels of mechanization.

CHAPTER X

The System Effects of Higher Mechanization[1]

CONTINUOUS MINING AS A NEW FRAME OF REFERENCE

The orthodox cutting longwall, which will be taken as a base for considering more highly mechanized systems, is itself a partially mechanized development of hand-got conveyor working, the getting and advancing phases being less mechanized than preparation. Unevenness of mechanization creates a technological disequilibrium in a work system which tends to have deleterious consequences in the socio-psychological dimension; these react back on the economic, and expected results are not obtained (*vide* p. 263).

The next stage in the mechanization of longwall systems has afforded a means of evening things up by extending to each of the other main tasks—filling, pulling, and stonework—some form of mechanical assistance. Such single task mechanization takes the form on the filling shift of the use, for example, of 'flight-loaders', a simple form of power loading. After the coal has been prepared by undercutting and shotfiring, some of the picks on the cutting machine are replaced with 'paddles' or 'flights' and the machine travels back along the face, the flights loading the coal on to the conveyor. Hand-pulling requires the belt to be broken into convenient lengths, rolled up, and moved over into the new track; in 'power-pulling' each half is wound round a prop at one end of the face by means of a rope attached to a small electric winch. On the stonework shift, hand-packing of gateway stone may be replaced by a machine-operated drag-line and scoop, which carries the shot stone into the goaf and rams it tight. Such

[1] Source papers: 6, 19.

machines are known as 'scraper-packers' or 'slushers'. A similar function is performed by 'crusher-stowers', which grind up the stone and stow it in the goaf under air pressure. The pulling and stonework main tasks are, however, only partially mechanized, for pullers still have to withdraw and reset chocks by hand and stonemen to drill and rip down the caunch. Nevertheless, the simultaneous application of such methods can appreciably raise the level of mechanization.

A further step is to use *multiple task machines*. These may combine two or more of the preparation tasks. Multi-jib cutters, for example, which replace drilling, cutting, scuffling, and shotfiring, may have two or three straight jibs which cut horizons into the coal, a curved jib for breaking down the top coal, and an automatic gummer for clearing the undercut. More highly developed machines combine preparation and getting. Several kinds are available which prepare the coal and load it on to a conveyor with a simple form of plough, eliminating hand-filling. These machines often run on an 'armoured' scraper-chain type of conveyor, which may be of the flexible or 'snaking' type. Snaking conveyors can be advanced to the coal face, as soon as the cutter-loader has passed, by hydraulic rams. In the most comprehensive systems now being attempted the armoured conveyor is coupled to the roof supports—the so-called 'self-advancing' props and chocks—with the result that preparation, getting, and advancing are so closely successional that conditions are given for *continuous* as opposed to *cyclical* mining. The advent of the flexible conveyor is as significant a step for the transition to continuous mining as was the original plate conveyor for the transition from single place orthodox longwall working.

Machines which simultaneously undertake preparation and getting fall into two main types: buttock and frontal getters. *Buttock getters* or cutter-loaders take the coal at right angles to the face to a depth of several feet. Having moved across the face, the machine is turned, and travels back taking another buttock; alternatively, it can be 'flitted' to its original starting point to begin again in the same direction. *Frontal getters*, which include the 'plough' and the 'scraper-box', work along the whole length of the face, moving it forward from end to end. Ploughs, of which several may be used simultaneously, have vertical and horizontal

FIGURE 6

SCHEMATIC DIAGRAM SHOWING DETAILS OF THE BASIC COAL-FACE THEIR PROPORTIONS OF PRODUCTIVE TIME, AND

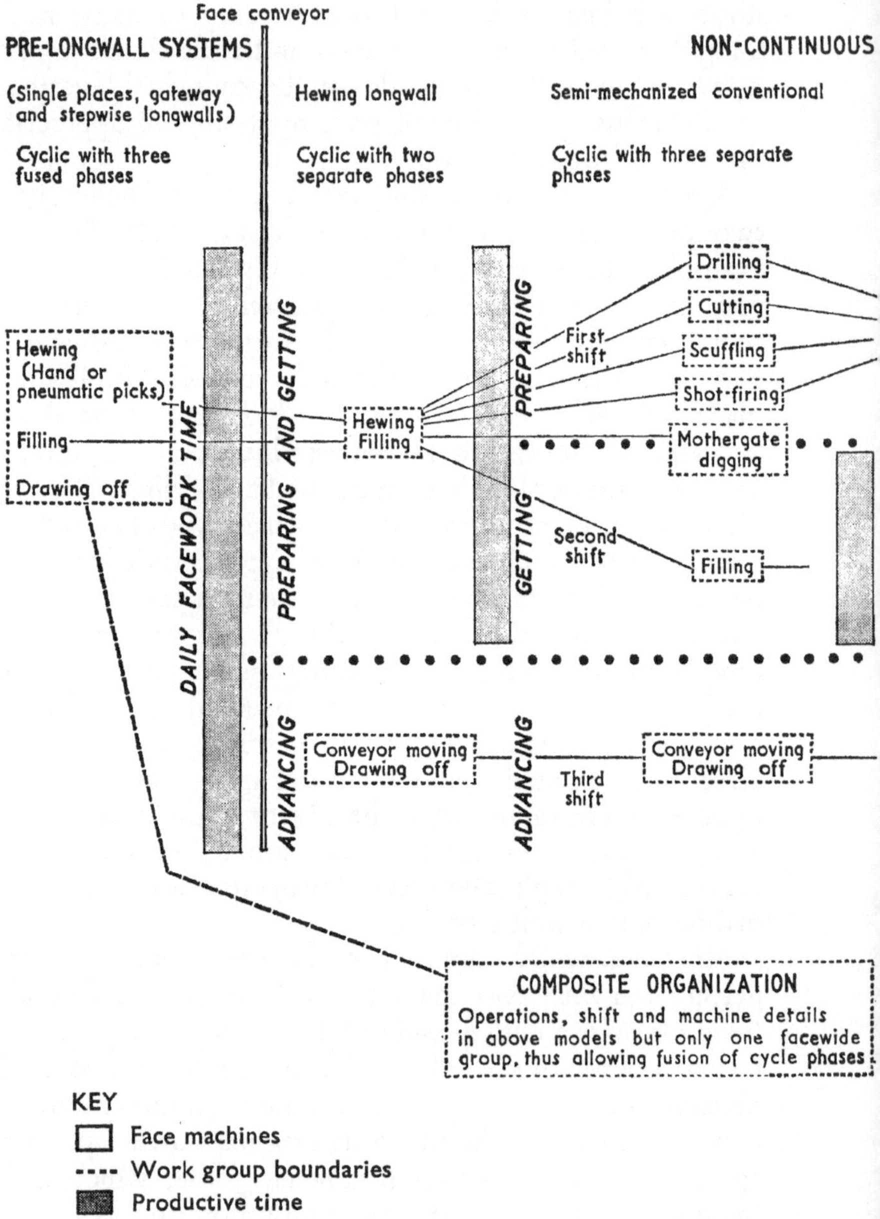

OPERATIONS IN SYSTEMS AT DIFFERENT LEVELS OF MECHANIZATION, THEIR ASSOCIATED TYPE OF SOCIAL ORGANIZATION

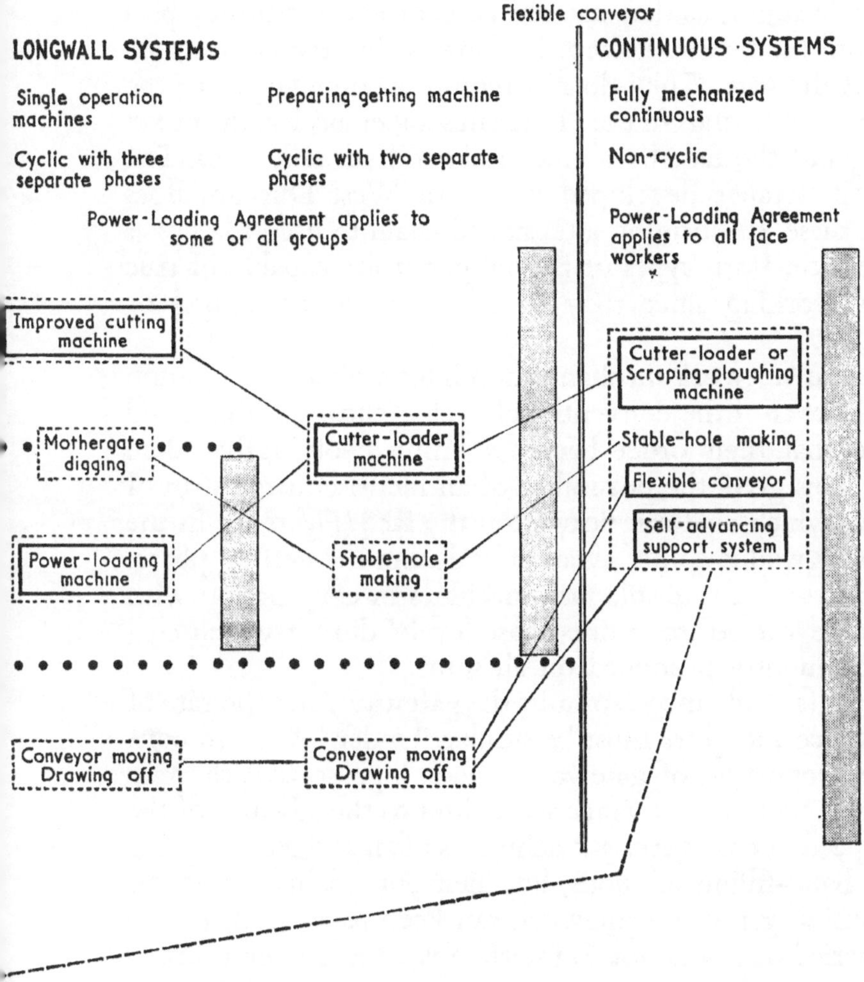

--- Indicates persistence of work organizations in which one team is responsible for all face operations (composite tradition)

•••• Phase boundaries

cutting blades which remove a thin slice of coal and guide it on to an armoured conveyor on which the plough is mounted. Scraper-boxes are simply boxes with cutting blades attached to their face-side edges; as they are drawn back and forth they peel the coal from the face and scrape it along within the boxes to be delivered at the gate. They do not require a conveyor since the boxes do the conveying. In the Haarman scraper-peeler, the boxes are held against the face by a guide rail and hydraulic rams. The tension-chain scraper developed in North West Durham does away with these, substituting a tensioned chain to hold the boxes against the face. Both types of frontal getter are capable of true continuous working since they do not need to be turned or flitted.

The trend in facework mechanization is towards a reintegration of the main tasks of the conventional cycle—cutting, filling, and pulling—so that their order becomes simultaneous rather than successive. This gives the possibility of an increase in the span of time during which coal can flow from the face (*Figure 6*). In the orthodox longwall it comes away only during the getting phase. The introduction of multiple task machines in conjunction with snaking conveyors creates a one-phase 'cycle' during which coal can be continuously produced on all shifts.

As the coal face advances, so must the gateways, and the rate of advance of face and gates must be similar. Under low seam conditions the proportion of gateway stone is high so that the rate at which stonework can advance sets a limit to the advance of the face. The speed of coal-getting machines is of an altogether higher order than hand-filling methods, but their potentiality cannot be exploited unless gateway stonework can keep pace—and, moreover, be carried out so as not to interfere with face operations.

There are a number of ways in which these limiting difficulties may be overcome, though, apart from the use of 'slushers', other known methods were not encountered in the pits visited in North West Durham. Changes in face layout are the most fundamental, as, for example, retreat, which was recommended in the Reid Report (1945). In this system roadways are first driven to the boundary; the face is then won out and worked back. Mining on retreat enables the nature of the 'ground' to be proven and this permits better choice regarding type and degree of mechanization.

Widespread on the European continent, this method is still the subject of controversy in Britain, where most longwalls have advancing faces. There are, however, other forms of layout in which facework and stonework can proceed simultaneously. In these coal is conveyed from the face in such a way as to avoid the area in the gate in which stonework is proceeding (Sheppard, 1951). They are equally applicable to conventional and to more highly mechanized systems. That such methods have not yet been more extensively tried is an indication of the degree to which a 'cycle' rather than a 'continuous' outlook is persisting.

THE CONTEXT OF SINGLE TASK MECHANIZATION

The immediate effect of single task machines, such as cutters, flight-loaders, power-pulling devices, and stone-slushers, is to raise the productivity of a particular task by taking away the hard manual work. Flight-loaders or stone-slushers, for example, lead to a reduction in the number of men on filling or gateway stonework. Alternatively, a flight-loader may permit a lengthening of the coal face and so an improvement in the coal-stone ratio; but the work to be done by the pullers is then increased. There is not, therefore, necessarily much reduction in the size of the face team as a whole, for men may be transferred from filling to pulling to cope with extra chocks or to put on face packs which sometimes become necessary on a longer face. The introduction of single task machines is to be thought of in terms of a redistribution of work roles rather than simply as a means of reducing team size.

Savings in manpower may, however, be fractional, as in power-pulling, in which the mechanized operation forms only a small part of the whole role. Whether such a saving can be realized depends on the form of work organization. On a conventionally organized longwall it is unlikely to result in any manpower reduction since the functional minimum for chock drawing (two) may remain the same. On a composite longwall the pullers are free to help the stonemen, and the net effect may permit a reduction in the number of men on pulling and stonework together. Composite organization permits more economical manning since size is not a simple addition of the number of men required for each task; fractional roles can be combined.

The introduction of single task machines changes the content of work roles and gives rise to a differentiation of activities among the task groups involved. The flight-loader gives the coal-getting phase (formerly filling) a structure similar to that of preparation, one or two of the men operating the machine and the others removing and resetting timber as it travels the face. The introduction of the machine converts an identical into a reciprocal role group. The slusher in stonework similarly creates differentiated activities—one man operating the winch, another supervising the packing, and a third acting as a communicating link between the two who may now be 30 yds apart.

Such changes are, of course, to be expected, but what was not always appreciated was the effect that single task machines might have on other activity groups on the same shift. On a conventional longwall, opening out a new area of the pit, it became necessary to 'solid stow' the face which was passing under property. Slushers were introduced with the aforementioned effect on the stonemen. The effect on the pullers was not anticipated. With no chocks to take and with gearheads moved up mechanically by the slusher winch, the role of puller was reduced to the one task of moving over the belt. Even with a reduction in the number of pullers to the functional minimum of two for the whole panel (a double unit with an unbroken face of 170 yds), pulling still remained a part-shift task. Its status with regard to cruciality for the cycle and its earning power were reduced. Having pulled the belt the two men concerned spent the rest of their time on shift work away from the face at a lower rate of pay. The barriers between the segregated task groups of the conventional pattern of organization prevented the natural adaptation to the new technology—a fusion of pulling with stonework.

Single task machines, especially in combination and more particularly when a flexible conveyor is used, create conditions where a change to continuous mining is but a small step. However, on the faces studied the qualitative changes resulting from use of single task machines had not created sufficient dissatisfaction with existing values and practices to stimulate a break-through to continuous mining. These machines had been fitted into partially mechanized cyclical longwall systems and their effect on this type of work organization was small. Where the prevailing outlook is towards continuous mining, partially mechanized longwalls can be adapted to continuous operation (Sheppard, 1951).

Though the more highly mechanized faces studied in the present research had achieved a steady state, their level of productivity was not impressively greater than that found on conventional longwalls and was substantially below the level theoretically possible. That the overall system may be in equilibrium does not in itself mean that single task mechanization will have much effect on production in a basically unchanged longwall organization, for the new equipment is assimilated into the conventional pattern. Power-loading, for example, when used purely as single task mechanization can reduce to one-fifth the time required to produce the same amount of coal, while the time required for preparation and advancing remains unchanged. The effect is to reduce overall cycle time—with less men but more dead-time—unless the face is lengthened and more loading time secured. Safely to lengthen the face often involves improving the method of roof control—by the use of more modern forms of chock or the deployment of additional labour on strip-packing. Technological change in one part of the system creates demands for changes in other parts. Similarly, where flexible conveyors are put into cyclical systems, without exploring the possibility of full concomitance in advancing and coal-getting, the production gains are disappointing, since a second coal-producing shift is not secured. Such points about system context have become matters of increasing comment in the Board's literature and in the mining press.

THE CONTEXT OF MULTIPLE TASK MECHANIZATION

Buttock getters include all combined preparation and getting systems where a work group travels the face with a machine—not only multi-jib cutter-loaders but installations where two ordinary coal cutters, one for undercutting and the other for flight-loading, are used together. Specific work roles vary, being to a not inconsiderable extent determined by the characteristics of particular machines. There are, however, four focal points which apply to all:

(a) the activity around the machine—operating the controls and removing and resetting supports
(b) the face-advancing activities—conveyor moving and advancing the chock line
(c) preparing stable-holes for turning the machine
(d) gateway stonework.

Nearly all buttock machines, at least until recently, have been used in some form of cyclical mining, full exploitation in continuous working being rare. The level of mechanization is usually to varying degrees uneven in that advancing is often unmechanized—a flexible conveyor may not be used and even where it is stonework may be done entirely by hand. The level of mechanization in the stable holes also varies, though shortwall cutting machines are sometimes used instead of pneumatic picks.

In moving buttock systems mechanized preparation and getting consists of two phases: the *face run*—the time during which the machine moves along the face and loads coal; and the *turn round*—the time involved in turning or flitting the machine, and in jibbing in and other activities required to prepare it for the next run. The time and amount of work involved in the second phase as compared with the first is a critical ratio. Where the turn round requires the work of several men for a whole shift, the system cannot free itself from a successive pattern and distinct groups perform different phases of one work cycle. On the other hand, a turn round requiring only a fraction of a shift gives very different possibilities. Provided the work is no more than the machine team can handle, one task group can undertake both preparation and getting. This is the case when a scraper-box or plough is used as a buttock getter, for turn round is almost immediate. The problem then arises of how soon the next coal-producing shift can be arranged—or how much dead-time must be tolerated or delay while gates are separately advanced.

Under high seam conditions buttock getters are the most common type of multiple task machine, but the size of the more complex models makes them unsuitable for low seam mining. No examples were available in North West Durham. A study was made, however, of such an installation in another Area, where a cutter-loader was being used in conjunction with a flexible conveyor. The turn round took an hour and a half, though the machine itself took no more than 20 minutes to prepare for the next run. The extra time was needed to complete chock-drawing and pack-building, which had been left unmechanized. The inability to keep the drawing-off and packing in step with the machine was responsible, against a criterion of ideal efficiency, for the loss of 70 tons of coal per shift—210 tons in the 24 hours, since three shifts were being attempted—and seven of these machines were operating with varying degrees of turn round delay. These losses could have been largely avoided, without undue cost, by increasing the manpower allocated to packing and drawing-off. But a major premise in developing the new system—under extreme pressure of the immediate employment situation—had been economy of manpower. To reverse this thinking when the system began to succeed proved a difficult reorientation and was only under first consideration when the research team was present, yet the addition of a few packers in order to gain machine time would have achieved a disproportionate gain in production.

The situation was worsened by the tension which existed between the machine crew—who had, as it were, taken possession of the new element and become the 'new people'—and the packers and drawers-off—who, being left with the old, had taken on an 'obsolescent identity'[1]. Their envy and quite bitter hostility disorganized their activities and slowed down their work pace. Being able to force the machine crew to help them out during the turn round was not without an element of unwitting and futile triumph. It was the findings of this case which first led to the hypothesis that unevenness in level of mechanization creates disequilibrium in a work system and that consequent disturbances in the socio-psychological dimension have economic repercussions in sub-standard results. Similar phenomena may be observed on numerous orthodox longwalls in the mutual attitudes of cutters and fillers. On one face undergoing extensive mechanization the stonemen complained to the research team of 'being forced' to do their work in the same way as their grandfathers and

[1] The authors are indebted to Floyd C. Mann, Programme Director of the Organizational Change Project of the Mental Health in Industry Programme, Institute of Social Research, University of Michigan, for communicating and discussing some of his unpublished findings on plant obsolescence.

made rather an aggressive display of a 'couldn't care less' attitude. The introduction some time later of slushers improved their morale; there were fewer grumblings from that time about the quality of the caunch work.

Frontal getters are more widespread in low seams and Haarman scraper-peelers were studied, both in Durham and in Holland, and a tension-chain scraper installation, locally developed, in Durham. In these installations usually one, at most two, shifts in the 24 hours were coal producing. Yet the scraper-box is well suited to low seam continuous mining in that it is mechanically simple, robust, and capable of standing long hours of work with little maintenance.

The three scraper faces—the Dutch Haarman, the Durham Haarman, and the tension-chain face—were technologically similar, but work organization differed in a number of respects. On all three, the machine team was an identical task group, the men on the face doing equal amounts of the same task. Difficulties are inherent in such a situation in that there is a tendency for those not having reciprocal tasks to lose cohesion. In fact, on all three faces powerful socio-psychological factors operated to counteract such a development. The Dutch system contained a status hierarchy among the faceworkers themselves which maintained integration through authoritarian discipline. The Durham faces had the tradition of marrow groups which ensured collaborative practices in the way facework was done; rather than keep to separate territories, men moved about and helped each other—with the roof, when the machine would crop some of the coal, or when the gate belt required attention.

Certain technical differences between the systems gave rise to specific differences in the content of the box-operating roles. On the Dutch face the method of roof control permitted the support line to advance by small amounts. This could go on right along the face and even get a little out of step without causing degeneration of the roof. Each man of the box-operating group could, therefore, be responsible not only for attending to the guide rail and rams but for advancing the roof supports in his own section. The tension-chain faces were supported by a chock line which had to be moved forward in a subsequent operation. Whereas the Dutch faceworker carried out all tasks, on the Durham faces there were two roles: one concerned with looking after the boxes and setting straps and props; the other, the orthodox pulling task of advancing the chock line. There was some overlap in work times in that the pulling team came on shift before the machine team had finished, and began advancing from the two outside ends of the double unit. Nevertheless, there were two distinct activity groups. These were not segregated single task groups, however, but part of a larger team rotating pulling and machining. Inasmuch as it allowed a change of role, and because pulling is a reciprocal task, the Durham pattern produced a more naturally cohesive organization than the Dutch, in which a strictly identical task group was held together by a status hierarchy.

In Durham, however, stonework was done by three entirely separate task and payment groups. This gave rise to dissatisfaction. Having no financial interest in the face, or role-relationship with those who worked there, the stonemen carried out their tasks to suit their own convenience rather than that of other groups. Tension disappeared when slushers were introduced, for the stonemen then also went on to the Power-Loading Agreement and the face team boundary was extended to include them. Similar difficulties arose on the tension-chain face through unmechanized stonework being done by a separate group. Unfinished caunches impeded the machine team who, however, put up with stone lying about, rather than finish off a task which was not their own. Nor did the machine team keep the stonemen in mind, though they often had spare time at the end of the shift which could have been used to prepare the caunches. The advantages of task continuity were lost.

On the Dutch Haarman, stonework in the two gateways was carried on while coal production was in progress—through a simple arrangement of lifting conveyors. Though this arrangement in itself would have permitted production to continue on a third shift, for other reasons, arising from the way the pit was organized, this became dead-time used for additional stonework. As the stonemen were on the same note as the machine team they took the coal from the gate headings and there were no separate hewing roles. No difficulties arose between the two stonework groups. They shared a bonus which included everybody between the face and the final loading point. All concerned had a positive interest in the efficient completion of the primary task and in achieving the co-ordination of their activities. In so far as two coal-producing shifts were secured the Dutch pattern was superior to the Durham.

This comparison of cases from widely different settings illustrated the related trends in higher mechanization towards composite organization and continuous production. The examples also show the constraining forces which prevent face projects from proceeding more than a limited distance along these lines when embedded in colliery systems based on cyclical assumptions.

CHAPTER XI

Organization and Manning under Full Mechanization[1]

No examples of mechanized faces were seen during the research which completely embodied the system characteristics that may be expected under these conditions when fully established. Nevertheless, certain principles concerning organization and manning began to become apparent which indicate the ways in which human resources may best be used in the technology that is coming into existence. These are set out in this chapter and are to be read in conjunction with the empirical studies of change projects described in Section Eight. The more advanced the technology, the more necessary does it become to specify the nature of the human factors and to design them into the configuration of the emergent socio-technical system, as is now being done in missile and electronic projects (Lyman, 1961).

INCREASED MAINTENANCE

The specific operating skills required on fully mechanized faces are not more difficult to acquire than those needed on partially mechanized longwalls. A *sine qua non*, however, is a higher level of general underground experience because of the greater variety of work situations in which a man is likely to find himself. The only exception arises with the more complex cutter-loaders where at least one member of the machine team, though not requiring to be a qualified fitter, needs sufficient knowledge to decide the required level of servicing should a fault develop. Such a man would make simple adjustments, but complex equipment depends on rapid attention from the specialist if production is not

[1] Source papers: 6, 19, 23.

to be interrupted. This raises for all mechanized faces such questions as the immediate replacement of damaged parts by spares, kept within close reach, while the former are taken away for repair, and the quickness with which tradesmen based on pit or Area workshops can be made available. The need is for a more elaborate and streamlined maintenance organization to keep more all-round miners in uninterrupted production.

ACTIVITY GROUPS WITH RECIPROCAL ROLES

A first basis for face organization is given by the various activity groups. In the case of frontal getters the distinctive processes are machining, advancing, gateway hewing, and gateway stonework. Hewing can be combined with stonework so that there are three essential teams, and, if the machine crew is organized as a marrow rather than a stint group, all activity groups can have reciprocal roles. The same conditions apply to buttock getters. The machine team consisting of one or two operators plus timberers is a reciprocal role group, as are also those in the stableholes.

COMPOSITE COMMITMENT

The machine team sets the pace of coal-getting and this in turn determines the work rate of all other groups, who are under continual pressure from something outside their own immediate activity with which they must keep up. As conditions are never fixed, there will be times when groups cannot keep pace with the machine without external help, so that whatever attachment men may feel to their own activity group, they must have an overriding loyalty to the work of the whole face, being prepared to move about and give support. This pattern requires not only a high level of general underground skill but systematic movement between groups to permit broadening of operating skills. It also requires a form of payment which induces a positive orientation to the efficient completion of the overall task. Separate paynotes for each group would be inappropriate and some form of common payment is desirable to embody in the most concrete form possible the commitment of the whole group to the face 'mission'.

LEADING FACEMEN

Maximum machine running time depends on rapid and effective ways of redeploying men and adapting activity to changing demands. In the composite longwall discussed in Chapter IX decision-taking and communication were diffused throughout the team. On a highly mechanized face the register of the production system is considerably higher and the reaction time of the regulating system must be correspondingly shorter. The composite pattern of self-regulation for hand-filling has too slow a reaction time for a fully mechanized face. A formal system of communication is needed with a decision-taking authority set up within the team. Since many decisions require immediate action and refer to specific operations, it is scarcely feasible for a deputy to undertake these functions—particularly under low seam conditions. In such circumstances a nominated or elected leader in each activity group can be made responsible. Such a man would be the link between his own team, other team leaders, and the management of the face. Roles of this kind can be rotated or permanent. In buttock systems where the cutter-loader determines the work rate of other groups present on the face at the same time, there would be a case for a role in the machine team carrying wider authority to co-ordinate their activities. The general tendency at present in handling the demands for tighter control in systems of higher mechanization is to increase the number of officials at the coal face, an overman as well as extra deputies being frequently assigned. The alternative is to build the required controls into the structure of the face team itself, so preserving its responsible autonomy and leaving to officials their proper task of managing the 'boundary conditions'.

CONCEPTUAL SKILL AND OPTIMUM MACHINE USE

Even with the most intricate machine a good deal of judgement must be exercised by its operator to ensure the best results. Moreover, as shown earlier, a machine working to full capacity will not ensure high output if other face groups are not in step. There is wide scope for the use of discretion (Jaques, 1956). This centres on the regulatory processes, because, if a machine is held up or works too far below capacity, the losses are altogether more serious than in lower register systems. A key regulating function is anticipation, which covers not only preventive maintenance of machinery and face conditions, but enters into the way all work is done in relation to other groups. When mechanical or geological difficulties call for rechannelling of effort or redeployment of men, the speed and efficiency with which counteraction is taken makes a significant difference to the duration of lost time. A high degree of anticipation coupled with early and effective counteraction maintain optimum machine use. This may not be maximum machine capacity, for at times it is more effective to maintain the rate of production below 100 per cent than to press for full speed and risk breakdown and disorganization. If in lower register systems the use of discretion is in no small measure a matter of how much physical effort is applied, under full mechanization it has a different quality, being much more a question of understanding the system as a whole and reacting with perspicacity to its demands. This involves what Emery and Oeser (1958) have called conceptual skill. More of the judgement of the full tradesman or professional is needed, in addition to the usual underground skills. The need for the exercise of conceptual skill is a reason for advocating the presence in each activity group of an unusually competent individual in the role of leading faceman.

REGULAR WAGES AND SALARIES

A regular wage which implies responsibility—the tradesman model—is a more appropriate form of reward under these circumstances than the piece-rate of the semi-skilled or unskilled operative. On the other hand, the tradition in the industry is that where production is higher, wages are higher. This tradition is recognized in the National Power-Loading Agreement, which combines a basic wage with a bonus component. Such a tradition has a social as well as an economic function. When a team works well, the greater pay arising is an easily comprehended index of 'pitman' status. As mechanization becomes more complete, however, this conception may well yield to one more in the direction of a salary-type contract. The level of a common basic wage, or salary, for faceworkers raises questions concerning the level of reward for officials whose tasks are also made more demanding of conceptual skill as mechanization increases, though fewer may be needed with the greater continuity and concentration of production. The effect of further mechanization is to raise the level of work in all roles. Unless this is ultimately recognized in the form and level of reward (Jaques, 1956), the more competent will seek to leave the industry and will not be replaced. The danger is that as a fully mechanized technology comes into being the supply will decrease of both men and managers of the competence required to obtain the results now possible.

PLANNED TEAMS

The traditional method of the self-selection of work teams which exists in the Durham coalfield strengthens internal rapport. It leads also to a grading of teams, the best workmen choosing each other. In single place working, this is wholly an advantage. The disadvantages of graded teams on partially mechanized longwalls become more pronounced with full mechanization. The delays and slow pace of a less able team handling a high capacity machine can lose hundreds of tons of coal. Cavilling is similarly out of place. Full integration of all groups is essential for mission commitment. Where conditions between faces vary, whole teams can be systematically rotated. The planned selection and allocation of teams on the basis of the whole seam population alone can ensure that key roles are always manned by men with the necessary conceptual skill, that inferior teams are not produced, and that there is an appropriate balance of skills between teams. This does not mean that all teams must be completely equal; the overall situation may well allow teams of varying capacities. It does mean that such differences should be planned. As self-selection of work teams is entrenched, and of special value in a high risk situation, personal choice can be allowed maximum scope within the limits imposed by planning—in which the lodge can share (cf. Chapter XXIX, p. 270).

A general effect of further mechanization is that there are fewer work roles—less narrowly specialized than formerly—and fewer activity groups—of a kind requiring more interlocking of the activities of different members. Each undertakes a larger portion of the cycle than any group on a conventional longwall, so that the problem of integration is less. Work roles share many of the characteristics of those associated with single place working and with partially mechanized composite systems. On the other hand, the reciprocal complexity of activity groups becomes greater and their interaction now requires more sensitive and channelled regulation in accordance with the higher tempo of production.

The account of coal face groups presented in the foregoing chapters suggests that conventional longwall organization is a divergent development between single place working and the more comprehensively mechanized systems now being introduced. At the level of partial mechanization, an alternative form of cycle group—the composite—consistent both with single place and more highly mechanized systems—had emerged and was operating more effectively than the conventional. Further mechanization gives rise to informal, if not always formal, change in work organization. How well any new organization matches the optimum requirement depends on how closely that previously existing resembles that now most desirable. Where further mechanization is intended, an easier changeover may be expected where the face has been composite, for this is likely to permit a more rapid transition to the type of work group best fitted to a more advanced technology.

PART II

COMPARATIVE STUDIES AND FIELD EXPERIMENTS

SECTIONS

SECTION FOUR

Comparative Studies of Performance and Control

CHAPTERS

CHAPTER XII

Evaluation of System Functioning[1]

The foregoing chapters have been concerned with the socio-psychological characteristics of the main systems of mining, with variations in the form of work organization within the same technology, and with the effects on work organization of further mechanization of the conventional longwall. Our next consideration is to develop measures that will describe these systems in operation and hence facilitate comparison between them. For reasons explained in Chapter II, direct comparison of production units was not possible, but during the course of fieldwork the notion was developed of comparing the various systems in terms of the degree to which they approached their potential limit of productivity. A method would, therefore, be required for estimating the difference between the actual productivity of the face and what it would be if those factors which interfered with its smooth operation could be eliminated.

The usual indicator of system functioning—output per man-shift—is too insensitive for this purpose. Its interpretation requires a knowledge of a great many factors implicitly taken into account by those in the industry when assessing o.m.s. trends—changes in seam height, the occurrence of faults, the duration of haulage breakdowns, etc. Many such factors, which are the essence of reality in the underground situation, are quantifiable and can be explicitly taken into account in comparing faces and systems. Two frames of reference which do so for cyclical systems have been constructed. The first assesses the effectiveness of the processes of cycle regulation in maintaining a steady state, the second is concerned with the outcome of cycle regulation and assesses production performance in relation to what is possible.

[1] Source papers: 1, 4, 9, 10, 13.

TABLE 7 CONCEPTS AND INDICATORS FOR THE EVALUATION OF SYSTEM FUNCTIONING

Cycle Regulation		*Production Performance*	
Concept	*Indicator*	*Concept*	*Indicator*
Cycle State	—Lag	Output Level	—Completeness
	—Advance		—Quality
Interference	—Internal	Task Situation	—Task Size
	—External		—Shift Length
Counteraction	—Substitution		—Work Rate
	—Overtime		—Reward
	—Reinforcement	Group Response	—Labour
			—Absence

CYCLE REGULATION

The objective of an analysis of cycle regulation is to assess the extent to which the pattern of cycle progress follows that scheduled; to determine the sources, degree, and consequences of interference which may disturb cycle progress; and to ascertain the nature and weight of counteractive measures applied to restore a steady state. Such an analysis, designed to identify the modal pattern of cycle regulation on a face, shows the probability of achieving normal cycle progress by the end of each shift under varying levels of interference and consequent counteraction.

The various tasks which comprise the cycle of operations may be isolated and distributed among the shifts on which they are scheduled to occur (cf. Chapter IX, *Figure 4* and *Table 5*). So long as the progress of the cycle is in keeping with the expected activity pattern, there is no *lag*, but, as soon as it starts to fall behind, lag begins to develop (cf. Chapter IX, *Table 6*). In some systems the cycle may get ahead of itself, i.e. be in *advance*. Objective indices have been developed to measure the *cycle state*: the length of face not cut or shotfired by the end of the cutting shift, the amount of coal not filled at the end of the filling shift, the number of chocks not drawn at the end of the pulling shift, etc., may indicate the amount of lag. In some cases these indicators may be quantitative measures; in others, because of the task structure, it is possible only to rate the amount of lag or advance at the end of each shift. In principle, therefore, it is possible to compare actual cycle progress with that scheduled.

The next stage is to determine for each phase of the cycle how much of the work done is concerned with the primary task of cycle completion and how much with non-cycle activity. The extent to which the activity pattern of the cycle is in any way disturbed, stopped, delayed, interrupted, etc., by technical breakdowns, waiting-on, and by shift work away from the face, must be determined. Such *interference* may be classified as *internal* when its source is within the face district, or *external* when it is beyond, and outside the control of the face group. Analysis of the sources and amount of interference provides one indicator of the kind and level of stress to which each shift group is exposed.

When the scheduled progress of the cycle is disturbed by interference, *counteraction* must be taken to re-establish a proper rate of cycle progress. Such action may be taken by the team on its own initiative, e.g. by increasing its work pace and by re-deploying itself. Where this does not happen or has insufficient effect, steps may have to be taken by management. Absentees may be replaced fully or partly by *substitutes*, *reinforcements* may be sent in for the whole or part of the shift and *overtime* may be worked. The reasons for overtime and reinforcement must also be ascertained. They may be necessary because of geological conditions, or because a shift group has been faced with developing cycle lag, or because time has had to be spent in making good deficiencies in the work of previous task groups. The addition or removal of labour from the face by management may under some conditions not be a direct consequence of conditions on that face. The optimization of operations in the seam as a whole may require that one face is reinforced at the expense of another—when a high producing face gets into difficulties men may be moved into it from a low producing face, even though the latter is producing regularly. The pattern of cycle regulation disclosed by analysis in the foregoing terms must, therefore, be interpreted in the light of conditions in the wider seam system and the 'style' and policy of seam management.

PRODUCTION PERFORMANCE

It is virtually impossible to combine in a single indicator all the variables relevant for an adequate characterization of performance; rather, a 'profile' of indicators is needed. The present frame of reference was initially developed for assessing the performance of composite longwalls and is particularly suited to that kind of organization, but in principle it is appropriate to conventional longwall organization and for the present descriptive purposes some generalization has been made. There are three main aspects of production performance, each of which comprises a number of indicators. The first is *output level*—actual related to potential production; the second is the *task situation*—the task to be carried out and the conditions under which the faces are operating; and the third, certain social and psychological aspects of the *group response* of the facework team to the task situation.

Output level relates the actual production of a face to that possible under ideal conditions. Under this heading come indicators of the *completeness* of production (e.g. the tonnage of coal obtained as a percentage of what could be expected from the seam section) and the *quality* of the output (e.g. the proportion of waste matter in the output from the face).

The quantifiable aspects of the task situation relate primarily to the size of the job facing the team, the amount of time and labour available to carry it out, and the reward that the system offers for complete production. The most readily available index of the *task size* is the weight of the seam (both coal and dirt) to be handled by the fillers, divided by the number of manshifts scheduled for this part of the task. The time available for work at the face is a net result of the normal *shift length* less time lost from interference with cycle operations and the time needed to travel to and from the coal face. The ratio of task size to shift length indicates the *rate of work* required of the fillers and the level of *work load stress*. Somewhat similar measures may be derived for the cutting shift but in the case of the pullers and stonemen a comparable measure is not so readily obtained because of the complex nature of their activities. Another aspect of the task situation may be subsumed under the heading of *reward*—the nature of the relationship between changes in the task size that faces the group and the possible level of earnings. For example, as shown in Chapter XIV, an increase in the amount of dirt band, together with a loss in coal height, considerably increases task size while reducing earnings, since wages are based exclusively on the amount of coal produced—unless 'consideration' is paid.

The effectiveness of the system is not to be assessed purely in terms of its success in completing the primary task. Account must also be taken of how far the team accomplishes the task with its assigned personnel resources—its ability to be *self-maintaining*. Overtime and reinforcements put into the face may be necessary to ensure cycle completion. On the other hand, when some members are absent, the team may still be able to complete its work without requiring substitutes. A *labour index*—the actual man hours spent at the face as a percentage of those scheduled—as well as being an indicator of economic cost, also reflects the extent to which the team is dependent upon resources outside itself for cycle completion. Although on occasion variation in the labour force supplied to a face may result from deliberate management action concerned with operations in the wider seam system, in general the index reflects the extent to which the team is able to mobilize its own resources and respond to the particular situation facing it. Another aspect of group response to the conditions under which the primary task has to be completed is the *pattern of absence*. Incidence rates for absences of varying duration, arising from *sickness*, *accident*, or '*no reason*', reflect the *social cost* of maintaining production as conditions—such as work load stress—change.[1]

The identification of relevant indicators of this kind and their interrelationship over an extended period of time provides a dynamic summary of the working of a system. Along with the modal pattern of cycle regulation, this permits diagnosis of operational difficulties which may have their roots in either the technological system or its associated work group organization. It is not, however, always possible to obtain such comprehensive data, but experience suggests that in order to summarize the

[1] For a detailed treatment of absence data as a withdrawal response to stress arising in the work environment *vide* Hill and Trist (1953, 1955).

characteristics of a system a basic minimum of three indicators is essential. These are the completeness of production index, the amount of external interference to which the system is exposed, and the labour index.

COMPARATIVE STUDIES

In single place working, faces were found to be operating close to their potential. Without changing the technology, improvement in productivity could be brought about only by training that would increase individual skill, or by replacing the less competent workers. In the case of longwall working, with its more complex technology and greater dependence on seam system functioning, there was much greater scope for improving the way the existing technology was used. The socio-psychological analysis of cutting longwall work organization suggested that the composite form was a better fit to the requirements of the mi ning system than the more widespread conventional form. To test this hypothesis, two comparative quantitative studies were made, applying, so far as the available data allowed, the concepts which have just been described. The faces studied were ordinary production units, geologically and technologically similar and not in any way experimental or otherwise atypical. The research design was based on a two-step comparison: the first, between two faces with widely differing sets of system characteristics, X and Y; the second, between two similar faces, Y and Y(X), one of which had some of the X system characteristics. If the first comparison showed that Y was superior to X and the second that Y was superior to Y(X), then the hypothesis would be sustained.

The first comparison was between a longwall organized on conventional lines and one organized on composite principles, faces representing the most extreme forms of work group organization encountered in the research. Comparison was made at a macroscopic level since there was reason to believe that differences of a major order would be disclosed. The second comparison was between two composite longwalls, one of which was less composite than the other, in the sense that its work organization had some features which were to be found on the conventional type of face. Since the productivity differences were expected to be small, comparison had to be carried out at a microscopic level. The next two chapters present the results of these comparative studies.

CHAPTER XIII

Face Team Organization and Maintaining Production[1]

CHARACTERISTICS OF TWO PANELS

In order to discover whether contrasting types of face group organization have any relationship to the level at which production is maintained, two longwalls—one organized on conventional and the other on composite lines—were compared. Both faces were in the same seam, at pits in the same neighbourhood. Geological conditions were alike as regards the nature of the coal, the type of roof and floor, and the dry and level character of the face areas, though minor differences existed in seam section and in amount of band (cf. p. 125). Both were double-unit panels with 80-yd faces either side of a mothergate. The same cutting technology was used and the haulage was similar. The primary task in both cases was the daily completion of a scheduled production cycle, and winding arrangements fixed the same relationship of tasks to shifts, with coal being filled off at specified times, alternating between foreshift and backshift.

The division of the primary task into the familiar sequence of cutting, scuffling, filling, hewing, drilling, pulling, and stonework meant that there was technical interdependence between the different main tasks. How any one task group carried out its activities affected, directly or indirectly, what had to be done by others. Similarly with the way places were kept and equipment handled or repaired. A good deal of co-ordination was required if the continuity necessary for the smooth running of the cycle was to be achieved.

The set of task roles were identical on both double units. There were, however, slight differences in manning, the total cycle

[1] Source papers: 2, 3, 4, 5, 11, 12, 13.

group on the conventional panel comprising 38 faceworkers, that on the composite 41[1]. These differences reflect the geological differences described. For the majority of roles—cutting, scuffling, drilling, and stonework—manning was identical. As, however, the fore caunch was carried further ahead, there was an extra hewer on the conventional panel, but one less filler on each face in view of the slightly lower coal height and the smaller amount of band. The composite faces each had two pairs of pullers, whereas each conventional face was advanced by one pulling group of three members. These differences reflect pit custom and also the somewhat increased task size of pulling on the composite panel.

The work force on the conventionally organized panel was divided into 14 separate groups, each on a different paynote. These groups were defined by their responsibility for one main task, which, drilling apart, was on only one face or in one gate. Beyond this territory and this activity they had neither responsibility nor reward. On the composite panel, the workers had formed themselves into one whole group on one equally shared paynote; all members were jointly responsible for all activities. Although manning the same set of roles, they had a system of rotation whereby they changed main tasks, shifts, and activity groups in a way they had prescribed for themselves (cf. Chapters XV to XVIII).

The effects on faceworker behaviour of these two different forms of organization will now be examined in terms of approach to work, non-cycle activity, inter-group relations, face experience, and absence. An assessment will then be made of their effects on cycle progress, regularity of production, and level of productivity. Finally, their consequence for management will be discussed, taking into account the roles of deputy, overman, and undermanager—those concerned with face and seam management.

DIFFERENCES IN FACEWORKER BEHAVIOUR

As regards *approach to work*, activities such as keeping the face in alignment and equipment in good running order, necessary to

[1] For full details regarding task roles, manning, etc., *vide* Chapter VII, p. 63, *Table 4* for the conventional panel and Chapter XV, p. 147, for the composite.

maintain the conditions for cycle completion, were of little concern to conventional work groups responsible only for one main task. Having no direct financial or group interest in the running of the cycle, they tended to be careless in these matters and not to mind how their way of doing their own main tasks might affect succeeding groups. Cuttermen, concentrating on yards cut, which is their basis of payment, did not bother if they cropped some of the coal and left it for the filler to dig out. Fillers, concentrating on tons filled, were not greatly worried by the consequences for the pullers of how they put in their supports. Pullers, in their turn, were not too careful about stacking withdrawn supports behind the belt and would leave them lying in the cutting track. All groups proceeded as though the cycle of operations were limited to their own task.

On the composite longwall, where there was only one team, all of whom shared a single primary task and a single paynote, groups anticipated the effects their activities might have on later shifts, and anything likely to cause extra work was avoided. The standard of workmanship, therefore, was higher. The face was squared off and completely cleared, with no band or coal left lying; timber was in a straight line, and gates and equipment were tidy. Quite different was the appearance of the conventional face, with spillage along the goafside, timber badly set, and gates and equipment in a rather neglected state.

TABLE 8 NON-CYCLE ACTIVITY ON A CONVENTIONAL LONGWALL

Task Group	*Percentage of Face Time*
Pullers	62[1]
Cuttermen	45
Scufflers	45
Driller	45
Fillers	37
Hewers	16
Stonemen	8
Whole Team	32

[1] Includes 43 per cent arising from coal filling

These different approaches to work give rise to differences in the proportion of time spent on activities necessary for the progress of the cycle. Ideally, all work done by a face team is on main tasks and certain sub-tasks essential to their performance. To be contrasted is non-productive ancillary work arising from disorganization or stoppages. Such *non-cycle activity* can never be entirely eliminated, but time spent waiting-on, doing overtime, or going on to shift work is an index of the extent to which the cycle is disturbed. *Table 8* sets out the proportion of face time spent on ancillary tasks by the various groups on the conventional longwall. One-third of all their activities was of such a kind, though the proportion varied considerably between different groups, as did the reasons for its occurrence. For the driller, cuttermen, and scufflers, time became available for shift work away from the face because their main tasks did not occupy a whole shift. Non-cycle activity for the fillers arose from interruptions caused by conveyor belt breakages and tub shortages. For the pullers it was increased beyond that of any other group by the call made on them to overcome cycle lag by filling off coal left on the face; an average of 22 per cent was left on by the fillers, and clearing this before beginning their own tasks accounted for 43 per cent of pullers' additional work. Only the hewers and stonemen, whose tasks on the whole were independent of those of other groups, had relatively small amounts of non-cycle activity.

For no task group, however, was more than 11 per cent of the time spent on ancillary work due to unavoidable causes. Over all groups this proportion was 7 per cent. The remaining 25 per cent was additional work made for one group by another. Although such extra work was seen as an imposition, there was no objection to its inheritance, because it was paid for, so that no one was discouraged from carrying out his main task in a way that created work for others. This pattern is referred to as the institution of *made work*, and is a latent effect of the division of the cycle aggregate into single task groups.

The common, equally shared paynote of the composite longwall was based on an inclusive fixed minimum, which covered sub- as well as main tasks and any ancillary work created. In addition, there was a large piece-rate component, 42 per cent of possible earnings being dependent on output. The men had an incentive, therefore, to complete the cycle without making unnecessary work. Main and sub-tasks accounted for virtually all time spent at the face, non-cycle activity being only half of one per cent.

Inter-group relations on the conventional longwall were at one and the same time competitive and collusive. Men had two sorts of relationship according to whether other faceworkers were inside or outside their marrow group. The marrow relationship, confined to members of their own main task group, was a close, friendly relationship, in which work and earnings were shared and members trusted and supported each other. But they had far fewer marrow than non-marrow relationships. These latter, which comprised their contacts with those in all other task groups, were competitive, suspicious, and unsupportive, with a psychological flavour of tension rather than ease, and offering opportunities for collusion rather than co-operation. The basis of the competition, which was covert, was ultimately financial, as each task group aimed to maximize its own earnings, while management aimed to hold total face costs within reasonable limits. There were, therefore, fourteen different pressure points on the same budget. But separate advantage could not be too openly sought without endangering 'worker solidarity', the traditional weapon against management. Collusion over made work provided a convenient way out of this dilemma, especially as it was largely unwitting. Men on the composite longwall had a common goal and only marrow relationships with all their fellow faceworkers. Problems of this kind could not arise.

TABLE 9 VARIETY OF WORK EXPERIENCE
(Averages for whole team)

Aspect of Work Experience	*Conventional Longwall*	*Composite Longwall*
Main Tasks worked at	1·0	3·6
Different Shifts worked on	2·0	2·9
Activity Groups worked with	1·0	5·5

Table 9 summarizes the main factors affecting day-to-day *experience of facework*. Men on the conventional longwall did their

one main task on only the two shifts to which it was assigned and always with the same group of marrows. On the composite longwall, because of the rotation system, the scope of day-to-day experience was much more varied. Men rotated between several main tasks, shared all three shifts, and moved from one activity group to another.

Facework places many stresses on the worker, particularly when things are not going well. One way of reducing these stresses is by making it possible for him to have a change of task, shift, or work place. When difficulties arise, one or two groups usually bear the brunt: if the roof is broken, the pullers and perhaps the fillers; if fragmentation in the gates is bad, the stonemen; if there is a small fault in the floor, the cutters. Seldom do all groups carry the burden of bad conditions equally. When difficulties occurred on the conventional longwall, the group with the extra load had no relief, but on the composite longwall the stress could be shared.

TABLE 10 ABSENCE RATES
(Per cent of possible shifts)

Reason for Absence	*Conventional Longwall*	*Composite Longwall*
No reason given	4·3	0·4
Sickness and other	8·9	4·6
Accident	6·8	3·2
Total	20·0	8·2

One effect of the differences in the scope of work experience and the possibility of sharing out the more stressful tasks can be seen in the *absence behaviour* of the two teams. Men on the composite longwall had in their task-shift rotation system a means of relief if some of the facework became unduly heavy, whereas on the conventional those who suffered from bad conditions had to put up with them. Their working life, therefore, was more stress-inducing, and the needs and temptations to withdraw—to be absent—were greater. This expectation of a higher level of absence is supported by the figures for comparative absence set out in *Table 10*, the rate for all reasons being higher for the conventional longwall, and that for voluntary absence over ten times as great.

TABLE 11 STATE OF CYCLE PROGRESS AT END OF FILLING SHIFT

(Per cent of cycles)

State of Cycle Progress	*Conventional Longwall*	*Composite Longwall*
In advance	0	22
Normal	31	73
Lagging	69	5
All cycles	100	100

EFFECTS ON PRODUCTION

We now turn to the effect of face group organization on maintaining production. *Cycle progress* on the conventional longwall tended to be erratic, since a good deal of time was lost on non-productive ancillary work caused by internal and external interference. During this time the cycle stood still. The best that could be hoped for was that main tasks would be up to schedule. For the cycle to be in advance was impossible, because, even if a group should finish its own task early, it could not go on to the next, as this was the preserve of another group. Indeed, the figures given in *Table 11* show that lag was usual on the conventional panel. There was a tendency for the fillers, especially, not to be able to finish so that management had to take counteractive measures—pay the pullers to complete the filling, and send reinforcements on to the face to complete the pulling. So usual was cycle lag on the conventional longwall that it required an average reinforcement of 6 per cent per week. On the composite longwall the cycle usually ran to schedule. It could and often did get ahead of itself. This was because of the task continuity which was practised, each shift group going on to the next tasks of the cycle as soon as they finished their own work. When lag did occur, the face team increased its pace of work in order to catch up, or at least to gain enough control so that the next cycle could proceed. They would, for example, under severe pressure, concentrate on finishing the most crucial tasks, leaving other work, such as that in the tailgates, to be completed later. In this way the composite longwall maintained itself without reinforcement. Counteraction was taken by the group itself, mainly through the practice of task continuity. This was something the formally segregated single task groups of the conventional face could not do. Only management was in a position to take counteraction.

As for *regularity of production* on the two faces, during the period of observation, the conventional longwall, with conditions quite normal, ran for only 12 weeks before losing a cut, and during these 12 weeks usually needed reinforcement to complete the cycle. The composite longwall, on the other hand, ran for 65 weeks without losing a cut, and at no time needed reinforcement.

TABLE 12 PRODUCTIVITY AS PERCENTAGE OF ESTIMATED FACE POTENTIAL

	Conventional Longwall	*Composite Longwall*
Without allowance for haulage system efficiency	67	95
With allowance	78	95

As regards *level of productivity*, the conventional longwall, in terms of output per manshift at the face, yielded 3·5 tons and the composite 5·3 tons. The conventional was perceived as at the norm for the conditions and the composite as above it.

The seam sections, however, were different: the conventional face averaged 21″ of coal and 1½″ of band, the composite 26″ of coal and 6″ of band, so that the latter had the advantage of more coal height together with the disadvantage of more band, while the conventional had the reverse conditions. The seam haulages, though similar in type, were not equally effective and the interference caused on the conventional face was greater. Comparison of the face o.m.s. was not therefore possible without adjustment of the figures, and each face was assessed against its own estimated potential. At 100 per cent efficiency 5·6 tons would have been expected from the composite panel and 4·5 tons from the conventional.

On this basis, and first without any allowance being made for the greater amount of interference arising from its less efficient haulage, the conventional longwall was working at 67 per cent of its potential (*Table 12*). In some measure this lesser efficiency of the haulage was due to poor maintenance resulting from back-bye labour being constantly drawn off to operating faces in order to cope with lagging cycles. It may therefore be regarded as a system defect. To some extent it was due also to the seam having been developed beyond the capacity of the haulage originally installed. A higher face productivity, however, would have increased the chances of something having been done about this when the working area was extended, so that once again the effects of system dysfunction cannot be entirely excluded.

If, however, in order to make the comparison more rigorous, full allowance is made for the higher level of external interference, the conventional face was working at 78 per cent of its potential. The composite, by contrast, was working at 95 per cent.

EFFECTS ON MANAGEMENT

Such different levels of effectiveness had very different consequences for face and seam management. On the conventional panel the entire burden of ensuring co-ordination of tasks and continuity of operations fell on officials. This entailed a great deal of effort, too much of which was expended on immediate measures necessary to counteract cycle dysfunction. With ancillary work at an overall level of 32 per cent, the deputies were heavily engaged in the detail of the ensuing complications, arguing with various task groups on precisely what needed doing and bargaining over amounts due and items eligible for payment. Apart from their statutory duties such as patrolling their districts, testing for gas and ensuring application of support rules, the time of deputies was almost entirely absorbed in taking emergency action over technical breakdowns or tub shortages, events arising from system dysfunctioning, and in administering an itemized price list (*vide* Chapter VII). Some idea of the demands of this latter activity can be gained from *Table 13*, which sets out the number of items involved in settling the pay of the different task groups on the conventional panel during an experimentally recorded quarter (13 weeks). Small opportunity was left to the deputies for attending to matters that, on a longer time span, would have reduced the level of interference.

TABLE 13

ITEMS IN THE PRICE LIST OF THE CONVENTIONAL PANEL DURING ONE QUARTER

A. Items for which the rate (though not the amount in any instance) is fixed by agreements

B. Items for which neither the rate nor the amount is fixed by agreements (items covered by 'shift work')

TASK GROUPS AFFECTED

Type of item	*Pullers*	*Fillers*	*Cutter-men*	*Scufflers*	*MG. Stonemen*	*Driller*	*Hewers*	*TG. Stonemen*	*All Groups*
A	15	11	4	3	5	4	6	4	52
B	22	16	18	18	14	14	9	7	118
A + B	37	27	22	21	19	18	15	11	170

The time and energy of the overmen were similarly consumed in dealing with immediate problems. With six to eight faces operating in the seam, the fore and back overmen were obliged to give priority to getting as much coal out as possible from whichever faces were filling off, and most of these were usually to some degree lagging. They deployed men and tubs accordingly, improvising to secure maximum production for any given day. The mastershifter coming in on the early nightshift had an even greater struggle, having first to eliminate whatever lag remained on all faces—with only limited winding time left—and then somehow to see that each face was advanced or cut so that the next day's cycles could proceed. What was intended as the principal maintenance and development shift became the principal trouble-shooting shift, with men drawn off from repair and development work to reinforce lagging faces and those tasks falling behind which alone could maintain the level of seam functioning. To break the vicious circle at overman level, however, was possible to no more than a limited extent, since the greater part of the dysfunctioning was being generated anew every day within the face districts themselves—25 per cent out of 32 per cent on the face selected for detailed study.

The undermanager, the first official with overall responsibility for the cycle, was already three steps in managerial rank away from the coal face, a distance too great to exercise immediate control. He was more worried about keeping down costs than about raising productivity, tacitly accepting the latter as impossible without a degree of change outside his scope to initiate. Such an attitude, expressing a solely defensive strategy, is the natural corollary of being in a situation where no positive improvement is seriously hoped for. In keeping with this attitude he saw his problems as arising more from the power of task groups to bid up prices than from the inflation of face costs by system dysfunctioning.

Both he and the colliery manager had commended the seam as an example of a 'very normal and well established' longwall operation—'a regular producer, pretty good conditions, a reasonable crowd of men though sharp about wages'—and this was its reputation also with higher management at Group and Area. The extent of cycle dysfunctioning was not perceived, the existing level of performance having come over the years to be accepted as the natural one. That the dysfunctioning might be due ultimately to the way the face team was organized was not believed when the present results were first discussed, though as time went on attitudes changed both at the colliery and in higher management. But initially an attempt was made to explain away as a special case what had been presented as typical. Several of the faces, it was said, were nearing the boundary so that only a limited investment in maintenance and new equipment had been justified; hopes were now placed in another group of faces soon to be opened out in a new area of the seam. When the research team visited these faces some months later, external interference was certainly reduced but coal was still being left on by the fillers, even if not so much or quite so often. The character of the disorganization, however, was unaltered and over the course of time in would in our opinion have built up again towards its former level, had not various technological changes ensued (*vide* Chapter XXX).

There was no greater contrast between the conventional and composite faces than in their management. As the composite organization was self-regulating, immediate cycle control was established by the group itself. The deputies needed neither to coerce, as it was in the interests of the men to get ahead, nor to bargain, as an allowance for an agreed range of sub- and ancillary tasks had been built into the Agreement. The comparable figure for the composite panel to the 170 items arising in the price list of the conventional panel was seven. Freed in this way, the deputies were able to give more time both to safety and to anticipating the needs of their districts. The centre of gravity of their role changed from 'propping up' a cycle always to some extent falling down on itself, to meeting the input and output needs of a going concern. To have the face cycle make demands on them rather than they on it was disconcerting at first and a number of deputies felt their jobs had vanished. All but the most rigid, however, were able to readjust by taking a more active part in regulating the interactions of the face and the seam systems, and to perceive the management of this 'boundary zone' as their real task.

The existence of a self-regulating primary work group exerts an upwards pressure in a managing system which affects all roles. With the elimination of made work by the face teams and with the deputies more active in seam liaison, one of the three overmen became superfluous. A single official, working a split shift, coped with both fore- and backshift, establishing unified control over the production shifts at two rather than three levels from the coal face. This emergence of unified production control over an unlagging cycle enabled the mastershifter to maximize the maintenance function of the nightshift so that almost all external interference was eliminated. The standard of maintenance in the entire seam system connected with these faces was of an altogether different order from that encountered elsewhere in the research.

All this allowed the undermanager to spread his attention to other seams which were more in need of it. He became more of an assistant manager. The extent to which a steady state had been reached may be gauged from the comment of the manager, 'I don't know now I have these faces in my pit'. At the opposite extreme is the degree of involvement of this same manager in the panel described in Section Six where the primary work group failed to become self-regulating. As the whole colliery was undergoing re-organization at this time, such involvement could be ill-afforded. The freedom needed higher up to manage change constructively is only won by establishing some freedom to manage at the bottom.

The emergence of a self-regulating primary work group undoes what Jaques (1951) has called the split at the bottom of the executive system, as there is no longer the same ultimate division into managers (of all ranks, including supervisors) and managed. Some of the managing has been taken over by the primary group —the part appropriate to its own task. Though this is what many in industry are allegedly seeking at the present time and though a managerial philosophy is coming into existence which makes this explicit (McGregor, 1960; Likert, 1961), such a development creates anxiety and produces resistance. In the present instance, the management-lodge negotiations went on for a year and might easily have broken down had not Area lent support. The first difficulty is over letting go the traditional managerial controls over the primary group; the second over accepting the challenge of the consequent rise in the level of work now required within management. To surmount these difficulties, however, is to replace job alienation in the worker by task-oriented commitment; thence, by reducing the pressure of immediate trouble-shooting, to increase the scope for creative problem-solving in management.

One qualification must be made to these conclusions. They have been drawn from the comparison of only one conventional and one composite face. The two selected faces were as closely similar as field conditions would allow and where dissimilarities existed adjustments have been introduced. The aim has been to approximate in a fieldwork situation the design of a crucial experiment, the efficacy of the comparison depending on the identity of conditions rather than the number of cases. Though complete identity cannot be claimed, the approximation obtained may be regarded as sufficient to establish the direction, if not the magnitude, of the result (cf. Lewin, 1935, Chapter I). There are, of course, many conventional faces operating more efficiently than that studied and other composite faces operating less well than the example given—and some of these are described in later chapters. Indeed, an overlap is to be expected, with the better conventional faces having production records superior to those of the less effective composite faces. One comparison does not enable the performance range of the two systems to be investigated, or estimates made of their mean levels of functional effectiveness. Such a qualification does not, however, invalidate the general conclusion concerning system characteristics: that the technical progress of the primary task is disrupted, in the conventional case, by disturbances induced by a fragmented social system; while, in the composite case, it is carried forward by the more continuous activity pattern arising from an integrated work group. This carrying forward of a production cycle by the social organization is equivalent at group level to the *traction* demonstrated by Baldamus (1961) as a positive characteristic of tasks at the level of the individual.

CHAPTER XIV

Work Load Stress and Cycle Regulation[1]

Placed on a 'scale of compositeness', the two longwalls compared in the previous chapter represent the extremes. A scale of compositeness would range from a strictly conventional organization with one-task—one-shift roles and no inter-change between task groups, to a fully composite organization with multi-task—multi-shift roles and completely free interchange between task groups. We shall now compare two composite longwalls, one of which—that used in the previous chapter—was closer to the composite end of the scale than the other.

Apart from the type of group organization which each developed for itself, these two panels were more alike than one would expect two longwalls to be even in the same seam in the same pit. They were adjacent. They used exactly the same technology and worked to the same agreement. Seam conditions were identical and the teams indistinguishable in qualifications and experience. They shared the same haulage and back-bye services. Both teams followed the composite work method as regards the practice of task continuity. The men were multi-skilled workmen, all being qualified in filling, drilling, pulling, and stonework, and between one-half and two-thirds also in cutting. Both teams were self-selected and accepted complete responsibility for allocating themselves to the various jobs that management required them to fill. The method of payment was an all-in flat rate plus a piece-rate bonus, the common paynote being equally divided in each case among all team members.

This was the general form of the composite system originating in the seam. Nevertheless, over time the panels developed rather different ways of organizing themselves. The main differences (summarized in *Table 14*) were as follows:

[1] Source papers: 8, 9, 10, 11, 12, 13.

(a) The group on No. 1 Panel organized itself as two face teams, each taking responsibility for manning the three shifts on its particular face. This face-wide organization distinguishes it from the panel-wide organization on No. 2 Panel, where no distinction between the two faces was made, the team dividing itself into two main shift groups, each of 20 men. Every fortnight the main shift groups alternated between the filling and pulling-stonework shifts and between them provided the men for the cutting shift, while the driller remained quasi-permanent.

(b) No. 1 Panel organized itself so that men tended to stick to one main task. For example, they would work as fillers or pullers, but not as both. In some ways this pattern is similar to that found on conventional longwalls, but since No. 1 Panel worked in the composite manner, all men became involved, in addition, in other tasks. Nevertheless, No. 1 Panel developed one-task roles. On No. 2 Panel, because men alternated between filling and pulling-stonework every fortnight and went on to cutting on a longer time basis, men carried out a range of different main tasks. They developed multi-task roles.

(c) On No. 1 Panel the team organized itself so that each specific job on the panel (e.g. stonework in the east tailgate on the

TABLE 14 DIFFERENCES BETWEEN PANELS IN FACE GROUP ORGANIZATION

No. 1 Panel (*Less composite organization*)	*No. 2 Panel* (*More composite organization*)
'Face-wide': organized as two rather separate face teams.	'Panel-wide': organized as two main alternating shift groups over the whole panel.
'One-task' jobs: men tend to work at only one main task.	'Multi-task' jobs: men rotate tasks systematically.
Each work place and task 'tied' to a particular man.	Work places and tasks not tied to individuals.
Not customary for men to move from one work group to another.	Men move freely from one work group to another.

backshift) was the responsibility of a particular individual. The men tied themselves to work places and tasks. On No. 2 Panel, so long as all work places and tasks were manned by qualified team members it was immaterial who they were. Jobs were not tied to individuals.

(d) No. 2 Panel team members moved freely from one activity group to another, not only from day to day, but from week to week. A man could, for example, work in a tailgate one day and in the fore caunch the next; or he might, when filling, change from one face to the other. On No. 1 Panel there was little movement of this kind; men definitely tended to stick to one work place. It was not their custom to move, though movement was permissible.

These four differences in face group organization were not so extreme as indicated, in that some of the features occurring on one panel were found on the other, especially as time went on (cf. Chapter XVIII). The differences nevertheless had consequences in three main areas—production performance, adaptation to changing conditions, and effectiveness of cycle regulation.

PRODUCTION PERFORMANCE

It was planned that the two panels should produce half of the total pit output with a fifth of the face manpower. From the outset, the panels achieved this target. A face o.m.s. of the order of 5·3 tons was maintained without reinforcement of the 41-man teams throughout the 20 months of their operation. Both panels went for over 15 months before losing a cycle. In all they lost only 12½ cycles out of 730 scheduled—1½ per cent.

Changes in conditions, however, must always be taken into account in assessing a performance record and to permit this the concept of a *production phase* was introduced to indicate a period of time during which conditions in the task environment remained relatively constant. The scheme of analysis described in Chapter XII was then applied to data taken from pit records to show what happened in four consecutive phases of production—A, B, C, and D—each of which lasted some 5 months and between which there were identifiable differences. *Figure 7* summarizes the results.

FIGURE 7

PERFORMANCE RECORD OF TWO COMPOSITE LONGWALLS

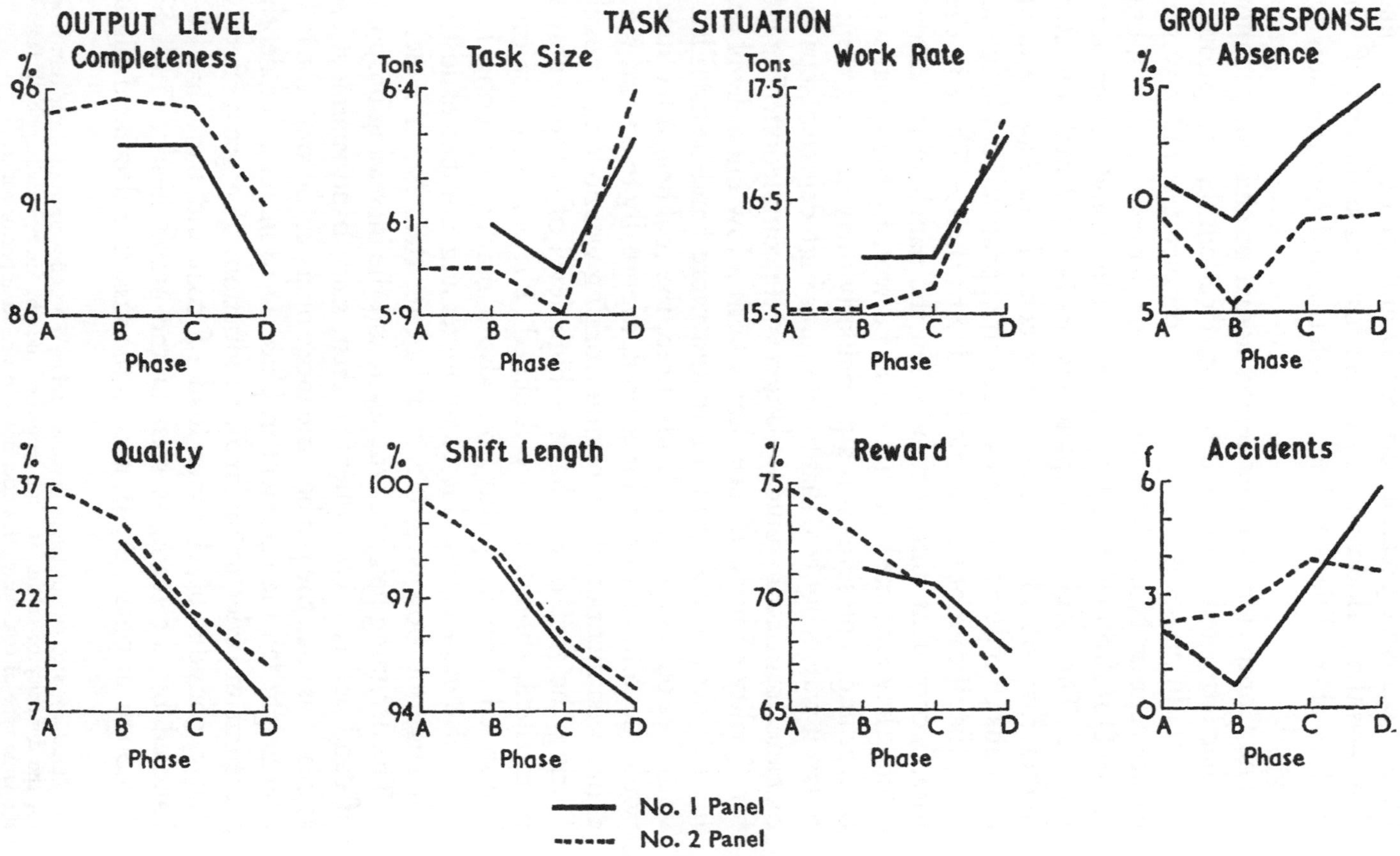

Differences were greatest between D and the other three phases. This was the period when cuts were lost and at the end of which geological conditions had so far deteriorated that the faces were stopped.

The Completeness index relates actual production per month to that estimated from coal height, face length, and amount of advance. In phases A, B, and C, the completeness index was over 93 per cent on No. 1 Panel and over 95 per cent on No. 2, but in phase D it dropped to 87 per cent on No. 1 and 90 per cent on No. 2.[1] The level of completeness on No. 2 Panel was always higher than on No. 1, by 2–3 per cent. In tonnage terms this amounts to only 20–30 tons a week. It appears so regularly, however, that it cannot be regarded as due to chance. There is, therefore, a consistent difference between the panels in the extent to which they extracted all that could be won by the given system of working. The seam was a particularly dirty one and from the outset it was recognized that to attempt a high degree of separation of band from coal would endanger cycle completion. The index for Quality of performance shows both a low initial level and a marked downward trend, more and more band being filled off with the coal. As the panels advanced, the coal height fell steadily while the amount of band increased, eventually to a point where, as has been mentioned, economic working was no longer possible. During the first three phases a high level of completeness was maintained, though quality declined. Phase D saw not only a continued drop in quality, but also a marked lowering in the level of completeness, due to an increasing number of lost cuts. Throughout, however, No. 2 did better than No. 1 Panel.

The declining height of the seam and the increasing proportion of band had two consequences. First, since band is almost twice as heavy as coal, the job became more onerous. Second, the falling coal height led management in phase D to increase the depth of undercut in order to maintain production. The gross effect was that a considerably heavier load of coal and band had to be handled by the team, as seen in the graph headed Task Size. Although in phases A, B, and C the task size decreased slightly,

[1] These levels of completeness become higher if an allowance of 4 per cent is made for the difficulty of cutting the full length of the jib. Since, however, both faces used the same type of coal-cutter this does not affect relative performance.

in phase D there was a very steep increase. As the panels advanced, the effective length of the shift available for work at the face decreased noticeably, the decrease shown in the graph headed Shift Length being equivalent, by the end of phase D, to a reduction of two manshifts per cycle—in a team of 41 men approximately 5 per cent. The operations of the cycle had therefore to be compressed into a shorter period and the graphs for Quality and Shift length follow a similar downward trend, showing that, with less time to do the job, quality suffered. With the increase in task size, and the decrease in time available, the teams had to work at a faster rate, as shown in the graph headed Work Rate. While the required rate was much the same in the first three phases, in phase D it increased greatly and for the fillers meant handling 17·5 tons per manshift. Relating this to the drop in completeness, one may conclude that beyond such a limit cycle completion becomes endangered.

On both panels the piece-work bonus was determined by the cubic yards of coal extracted. As coal height became less, so did possible earnings, since at that time the piece-rate did not take changes in coal height into account. Another way of looking at this situation is to consider what proportion of the effort expended during the shift was devoted to filling off coal as distinct from band. If 10 tons of coal and band had to be handled and 8 tons of this was coal, then 80 per cent of the effort would be rewarded; on the other hand, if there were 12 tons of coal and band of which only 8 tons was coal, then no more than 66 per cent of a man's work would be paid for. The graph headed Reward shows that, as the panels advanced, possible earnings bore less and less relation to the effort required. The similarity between the graphs for Quality and Reward shows one effect of this.

Though No. 2 had a rather rougher time than No. 1 Panel, both followed a similar course, completeness of production falling sharply in the last phase when cycles were lost. This coincided with an increase in task size, a reduction in shift length, a faster work rate, and a growing disparity between effort and reward. All these changes operated as a stress on the team, who had, essentially, to deal with a bigger load. This may be referred to as *work load stress*.

ADAPTATION TO CHANGING CONDITIONS

One useful indicator of response to the stress of increased work load is provided by the absence record of the panels. During the period of the study, the faceworker absence rate for the pit as a whole averaged 12 per cent, but the panels themselves were below this level, 10 per cent for No. 1 and 8 per cent for No. 2. The differences are significant not only between both panels and the rest of the pit—with its mixture of conventional longwalls, composite shortwalls, and single places—but between the two panels themselves. Again the trend favours No. 2. To investigate this more fully, absences during phase D—when the work load increased—have been compared with those in phase C in which the work load was much the same as during the preceding phases.[1]

The changes between these two phases produced striking differences of response from the two panels, as may be seen on the graphs headed Absence and Accidents. On No. 1 the absence rate increased with rising stress, on No. 2 Panel it remained unchanged. *Figure 8* presents a fuller picture of these changes under Level of Stress and Withdrawal (which gives the incidence trend in absences from all sources). On No. 1 Panel the rising absence rate was largely due to an increase in absences lasting only one day.[2] There was, particularly, an increase in the incidence of one-day sicknesses, and also an increase in accidents leading to a day off. No significant change, however, occurred in voluntary ('no reason') absences. On No. 2 Panel, the changes in one-day absences were so small that they could have come about by chance. While increased stress showed itself on No. 1 Panel as a definite increase in the number of single days off, the incidence in phase D being double that in C, No. 2 Panel showed no difference. It follows that the team must have had some alternative and more effective way of coping with work load stress.

Given the equivalence of the two groups, the differences cannot be explained by assuming that the men on one panel were more susceptible to infection or more accident prone than those on the

[1] The drop in the absence figures on both panels in phase B was due to seasonal factors, but these do not affect the comparison of C and D.

[2] cf. Hill and Trist (1953, 1955) regarding the socio-psychological significance of one-day absences.

FIGURE 8

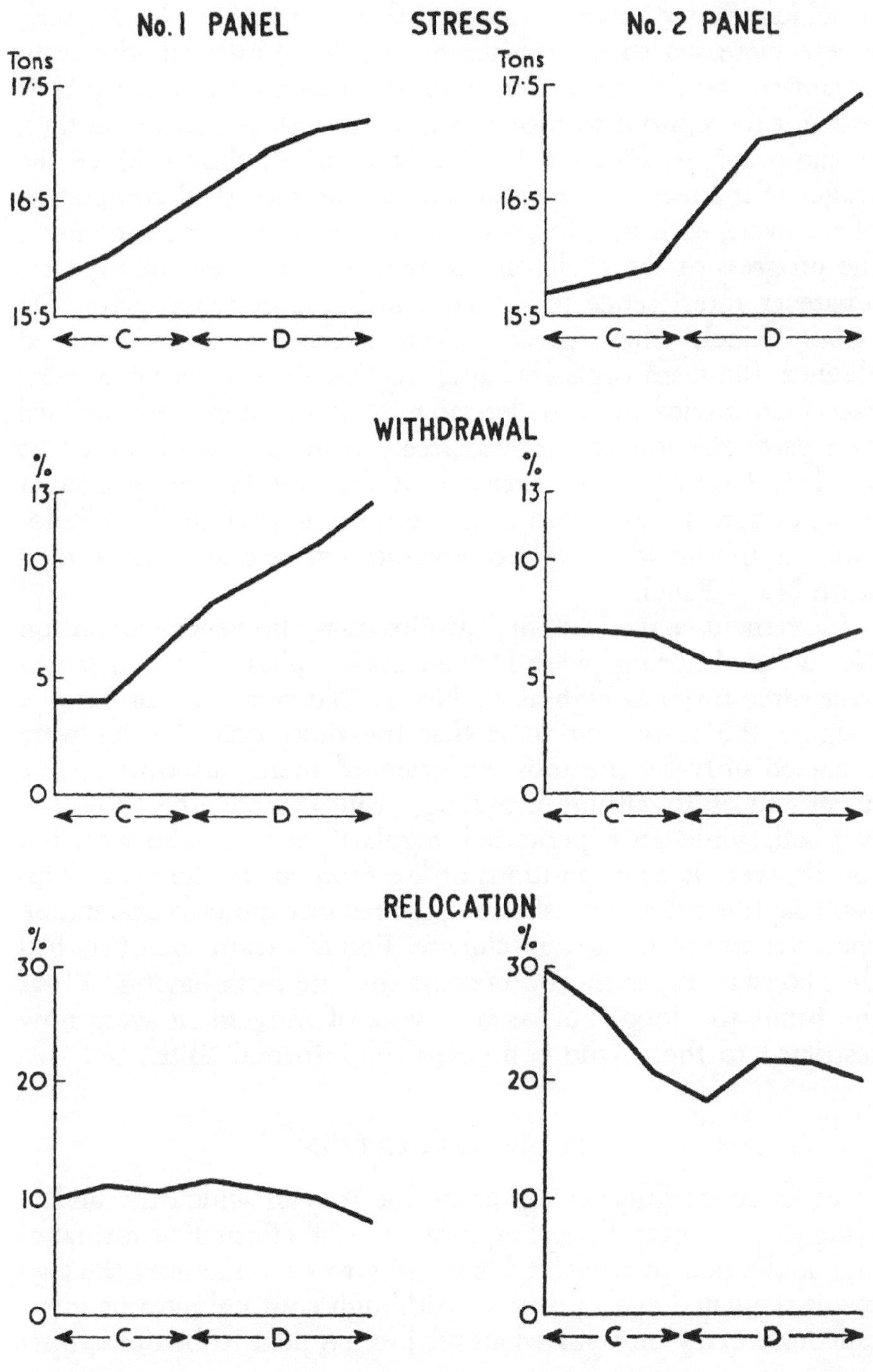
No. 1 PANEL
STRESS
No. 2 PANEL
Tons
17·5
16·5
15·5
C
D
WITHDRAWAL
%
13
10
5
0
RELOCATION
30
20
10

other. Rather, an explanation must be sought in terms of the way in which the two teams organized their work. On No. 1 Panel, where increased stress and absence go hand-in-hand, the team organized itself so that each man was tied to one main job. It was not the custom to move from one work group to another, as the graph in *Figure 8* headed Relocation, shows. Since the wages of the team were dependent on the successful completion of the cycle, each man felt personally responsible for maintaining the progress of the cycle on his own shift and for coping with whatever interference might arise in his own work place. On No. 2 Panel, where greater stress did not lead to increased absence, the team organized itself so that over a period of time each man carried out a wider range of tasks; men were not tied to a particular job and moved freely from one work group to another. Consequently, excess load did not fall on particular men; rather, it was spread over the team as a whole. The Relocation graph shows a level of movement more than twice as high as on No. 1 Panel.

Movement across activity groups was, however, lower on No. 2 Panel during phase D than during phase C when it was over three times as high as on No. 1. When conditions became rougher the team saw to it that the most crucial roles were occupied only by the most experienced men. Substitutes were never sent on to cutting or pulling, or into a gate where the roof was bad, while less experienced 'regulars', or men who were not too fit, were kept in positions of less moment to the cycle. This was adaptive behaviour, showing the realism and task-orientation characteristic of the group climate. Enough team members had the necessary experience to prevent any one from having to bear the brunt too long, but certain types of movement were now restricted to those who composed the informal 'élite'.

CYCLE REGULATION

In order more fully to appreciate the way in which the underlying differences in face group organization affected performance and adaptation to stress, it is necessary to examine how the two teams regulated cycle progress. Although composite teams spontaneously carry on with whatever job has next to be done, how far a shift can, or even should, proceed with the work of the cycle is governed by a very complex set of factors; basically, it depends on what stage the cycle is at when the men come on shift. A quantitative study of the regulation of cycle progress during phase D, when the roughest conditions were experienced, was made by comparing, for each different beginning, the average state of the cycle at the end of the shift. For example, when the cycle is lagging through the cuttermen not having finished their work, the fillers manage to finish the cutting and also to complete their own work—they put a spurt on in order to eliminate the lag. When the cycle is normal, normal progress is made. When the cuttermen achieve a slight or medium advance and give the fillers a start, by the end of the shift the fillers have pushed the cycle a little further ahead. When, however, substantial advance is made by the cuttermen, the fillers aim simply to maintain it, there being no virtue in the team getting the cycle too far ahead, for the smooth running of the seam as a whole could be disrupted. The inference to be drawn for all shifts—cutting, filling, pulling, and stonework—is that the teams were able to regulate their work to suit the varying conditions and to satisfy the requirements for optimum running of the seam system as a whole.

The panels differed considerably in their method of gaining control over cycle lag. On No. 1 Panel, lag of whatever degree—short of actual breakdown of the cycle—was eliminated during the shift that inherited it, whereas on No. 2 Panel two or three shifts would be allowed to elapse before the cycle was brought back into phase. The men on No. 1 Panel would not pass on any inherited lag to their marrows on the succeeding shift. Their attitude was that every shift should attempt to bring the cycle back into phase, regardless of whether control could be more economically achieved by passing on some of their work to the next shift. When the work load increased, each group, by attempting complete control over any lag it might inherit as well as aiming to finish its own job, raised still further the level of stress. In time, the greater strain which men experienced resulted in greater absence. Such were the consequences of a face group organization which tied men to particular jobs and limited interchange between work groups. On No. 2 Panel, by systematically rotating the various shifts, men came to know better what could be done under the conditions of each. They did not expect a particular shift to achieve complete control but accepted as quite reasonable that some of the consequences should be coped with by later shifts. With a span of three shifts, as compared with one in which to eliminate lag, they therefore experienced less strain and no significant increase in absence occurred.

There was one other difference between the panels. To ensure a smooth succession of cycles, operations on the leading face needed to be slightly ahead of those on the lagging face and close co-operation of the men working on the two faces was necessary for this optimum situation to be achieved. No. 2 Panel, which was organized on a panel-wide basis, always kept operations on its leading face slightly ahead of those on the lagging face. When anyone was required for shift work in the gates, men were drawn from the lagging rather than the leading face. The team also concentrated lost cuts on the lagging face—making the best of a bad job. No. 1 Panel, with its two rather separate face teams, operated quite differently. They kept operations on both the leading and lagging faces closely in step. When men were required for shift work away from the face, they were drawn equally from both faces. Cuts were lost with the same frequency on both. This overall method of cycle regulation was sub-optimum for the particular technology and the double-unit layout. The difference between panel-wide and face-wide organization had very real consequences for the regulation and progress of the cycle.

This comparative study indicates that one form of face group organization was a better fit than the other to the requirements of the situation:

(a) in terms of the productivity of the faces—though there was little to choose between them, such differences as there were consistently favoured the group organized on a panel-wide basis

(b) in terms of regulating cycle progress—though on both panels the practice of task continuity enabled the teams to get sufficiently ahead to cope with inevitable and unpredictable difficulties and interferences, the differences definitely favoured the same panel

(c) in terms of the social cost of maintaining a high production record—in sickness and accident absences, which arise from the way increased work load stress is coped with, there were considerable differences, which again favoured the same panel.

The face group organization which was panel-wide, embodied systematic rotation of the various jobs among team members, and did not tie a man to only one job, work group or face, was the more effective in maintaining the smooth flow of the cycle and in coping with increased work load stress.

The differences in the operational records of these two composite longwalls are to be accounted for by the presence of certain 'conventional' features in the face group organization of the less effective panel. The comparison of a conventional and a composite longwall presented in the previous chapter showed that the superior production performance of composite organization stems from its more effective regulation of cycle progress. The second comparison, presented in this chapter, is more stringent—that of two composite longwalls working under almost identical conditions though differing in their internal work organization. The results show unequivocally that the presence of conventional characteristics affected the way one of the panels regulated its work so as to depress its performance level, while increasing 'casualties' in the face team. This two-step comparison leads to the general conclusion that, for workers carrying out a primary task comprising interdependent component activities interchangeable between group members, the composite form of organization has inherent characteristics more conducive to productive effectiveness, work satisfaction, and social health than that based on separately treated single task groups.

SECTION FIVE

The Creativeness of Composite Work Groups

CHAPTERS

CHAPTER XV

Origin and Formation of Two Composite Teams[1]

This study of the social creativeness of autonomous work groups is based on observation and interviews at the coal face and on an analysis of reports and records covering the life of the two double-unit Manley panels described in Chapter XIV. Each consisted of 80-yd faces east and west of a mothergate. They were both advancing northward from the same trunk road and were adjacent, No. 1 Panel being to the west of No. 2. Despite their proximity, they developed on rather different lines.

Each panel had the same manpower complement of 41 men, allocated to the following roles:

4 cuttermen	2 to each face
2 scufflers	1 to each face
14 fillers	7 to each face
2 hewers	in the mothergate
8 pullers	2 pairs to each face
10 stonemen	2 to each tailgate, 3 to the mothergate fore caunch, and 3 to the back caunch
1 driller	for the whole panel.

All men were qualified to undertake any task except that initially 16 men on No. 1 Panel and 15 men on No. 2 Panel were not formally qualified to cut.

It was pit policy that filling should be done on either foreshift or backshift, the other parts of the cycle varying accordingly:

Cutting:	Night	or	Fore
Filling:	Fore	,,	Back
Drilling (split between):	Fore and Back	,,	Back and Night
Pulling and Stonework:	Back	,,	Night

[1] Source papers: 7, 21.

The capacity of the haulage required that when one panel filled on the foreshift, the other filled on the backshift.

No. 2 Panel, which organized itself as a whole, was the first to start on 11 October 1954, and ran for six weeks before the East face of No. 1 started, the West face following 13 weeks later. Each of the three teams made changes in its system of rotating tasks and shifts, so that there were a number of *rotation system phases*. The times at which these changes occurred are set out in *Figure 9*. The 81 working weeks covered have been divided into *periods* of four weeks.

The management-lodge discussions which led to the formation of the teams were concerned, *inter alia*, with the allocation of men to tasks and shifts. Management wanted men to change their *tasks* at not less than four-weekly intervals in order to minimize the loss of working efficiency said to accompany change from one task to another. The lodge, concerned to avoid the longwall dread of men being permanently on a 'bad' shift, wanted a change of *shift* every week. A compromise of coincidental changes of both task and shift every fortnight was finally agreed. The panel teams were to be wholly responsible for allocating a man to every task on every shift and for ensuring that there was adequate rotation of men to different shifts. The task-shift rotation systems that so emerged were therefore developments entirely internal to the panel groups and their interaction with each other.

The initiative in making up the team and in devising the system of task-shift rotation for No. 2 Panel was taken by two men, both of whom had taken part in the management-lodge discussions as members of the lodge committee. They therefore formed a direct link between the negotiating group and the panel team and had first-hand knowledge of the intention that men should change their tasks and shifts every two weeks. On No. 1 Panel, neither the East nor the West face team had any such direct contact through common membership of the negotiating group. Their knowledge of the intentions regarding rotation was at best second-hand.

The similarities and differences that have developed between the task-shift rotation systems of the panels and faces must also be seen against the background of the experience of the men who

FIGURE 9

THE DURATION OF ROTATION SYSTEM PHASES

(Figures in brackets indicate the week in which the phase commenced)

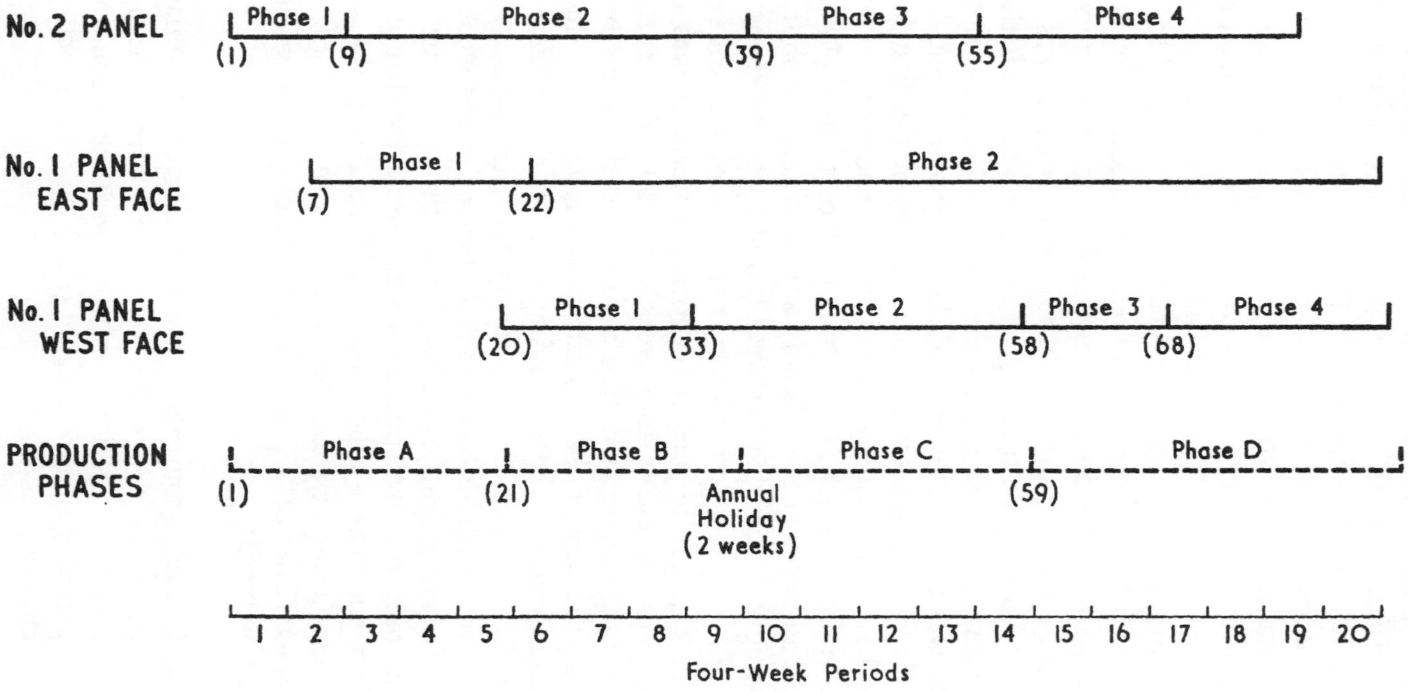

made up the teams. All members of both teams shared a common experience of working in composite shortwall units and many of them had worked together in the same team before going on to the longwalls. The team that made up No. 2 Panel came from four existing sets: 20 from a set of 23, 13 from a set of 20, and two groups of four from 15-man teams. They had, therefore, in the first two large sub-groups especially, recent experience of each other's task and shift preferences and skills. No. 1 Panel was made up from three 15-man teams (less four older men who did not want to go on the longwalls). The team, however, never went on to the panel as a group of 41 men. It was cavilled into two groups, 23 men taking up the East face, while the remaining 18 worked on a shortwall elsewhere in the seam until the West face was ready. The original shortwall teams were broken up at random, with the result that members of the East face team did not have such widespread or recent experience of each other as did those of either No. 2 Panel or No. 1 Panel (West).

In addition to the officially agreed change of shift times every fortnight, other changes were found to occur in the intervening weeks (*vide* Chapter XVI, p. 157). In order to discover how far these changes formed a recurring pattern, it was necessary to compare successive four-weekly periods. *Figure 10* shows the percentage of times during each period that the same men were doing the same tasks or shifts in any given week that they were doing during the corresponding week of the previous period; the task (or shift) of man X during the first week of period 4 is compared with his task (or shift) during the first week of period 3, etc. A man might, of course, be on the same shift or doing the same task in corresponding weeks purely by chance; in this case 26–28 per cent of the men would be doing the same task in corresponding weeks. From *Figure 10* some of the general characteristics of the task-shift rotation systems which developed may be listed:

(a) In each group there is a high degree of similarity in both the tasks and shifts worked in corresponding weeks of successive periods, which (with the one exception of period 7 of No. 1 Panel (East)) is very much greater than chance.

FIGURE 10

PERCENTAGE SIMILARITY BETWEEN TASKS AND SHIFTS WORKED IN CORRESPONDING WEEKS OF SUCCESSIVE PERIODS

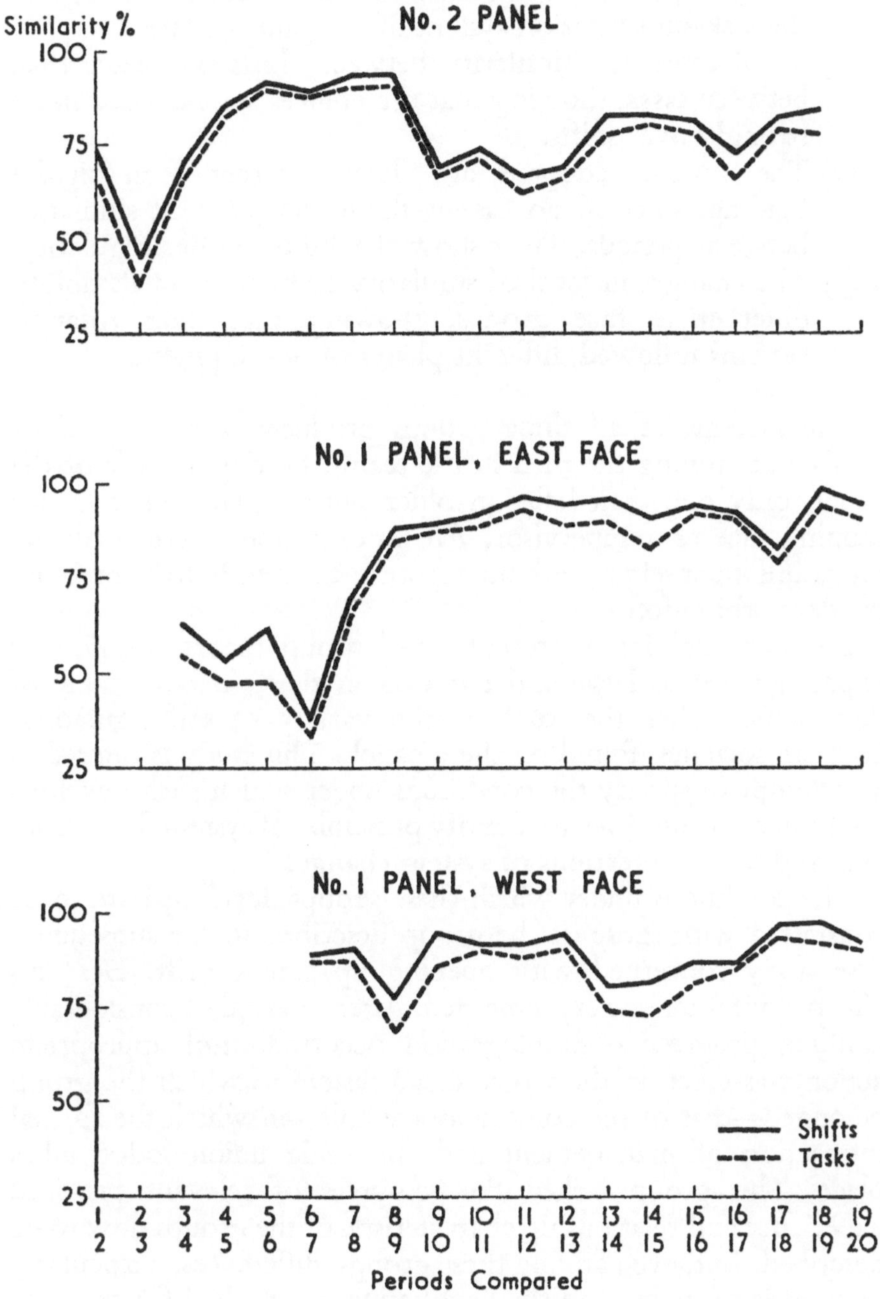

(b) Where there is a marked change in the level of similarity between periods, this coincides with a known change in the task-shift rotation system of the panel or face.
(c) In all cases, the similarity between shifts is greater than between tasks, showing that preferences are exercised more strongly over shifts.
(d) The systems adopted are flexible rather than rigidly binding, since in no case is there 100 per cent similarity between periods. They show flexibility within regularity.
(e) The changes in level of similarity and degree of flexibility differentiate face groups, showing that their rotation systems followed different phases of development.

The systems of all three groups produced a high level of satisfaction among the men in the teams. During the life of the panels only one man left—an older worker who went to the training face as a supervisor. All modifications were made by the teams themselves without recourse to outside help or independent arbitration.

Prevalent opinion in recent small group theory would not expect groups so large and complex as these to be capable of developing either the kind or the variety of self-regulating, internal 'societies' found on these panels. This study is offered as an attempt to specify the conditions under which such developments become possible, to identify principles of system formation, and to discover directions of system change.

The conditions under which these groups developed are to be contrasted with those of the group described in the subsequent case study concerned with opening up a new drift. Here no effective internal 'society' emerged. After a period of considerable conflict, progressive disintegration occurred until appropriate action was taken in the larger social system to which the group belonged—that of the colliery as a whole—in which the formal interaction of management and the trade union lodge takes place. This group lacked the resources of relevant previous experience and relationship characteristic of the groups now to be described. But, even among these groups, differences, particularly as regards relatedness to the negotiating group, had far-reaching consequences. The series of faces discussed in Chapters XXV and

XXVI, dealing with the change from conventional to composite working in another seam, constitute an intervening set between the drift and the Manley panels. The total range indicates the different outcomes to be expected, as regards effective internal development, when different socio-psychological conditions obtain in the 'life-space' of working groups.

CHAPTER XVI

Contrasting Patterns of Initial Deployment[1]

PANEL-WIDE SHIFT EXCHANGE (NO. 2 PANEL, PHASES 1 AND 2)

Phase 1 of No. 2 Panel covers the first eight weeks of its life. For the first six it was the only unit and filled regularly on the backshift. When, however, the East face of No. 1 Panel started in week 7, it filled on the foreshift, so that No. 2 Panel did not need to adjust to alternate filling times until week 9, which marks the beginning of its second phase.

The 41 men who came together to form No. 2 Panel had to provide six cuttermen on the foreshift, 16 fillers on the backshift, 18 men for pulling and stonework on the nightshift, and one driller. They had also to ensure that every man changed his task and shift every fortnight. Their solution was to organize themselves into two main alternating 'shift' groups of 20, round the two men who had taken part in the negotiations. From the start one man—whoever occupied the role of driller—was not a member of either group. During any fortnight one of these two groups would supply the 16 men for the filling-hewing shift, while the other supplied the 18 for pulling-stonework. The four remaining from the first and the two from the second 20 came together as the six required on the cutting shift, and were regarded as 'lent out' from their basic groups (*Figure 11*).

This model approximates closely to the intention of the initial discussions that all men should have a change of task and shift every two weeks. The exception concerns drilling, which was done by only one man, who overlapped filling and pulling-stonework. This was a personal preference.

When the panel started, the team captains arranged the dis-

[1] Source paper: 7.

FIGURE 11
ROTATION SYSTEM OF NO. 2 PANEL, PHASE I

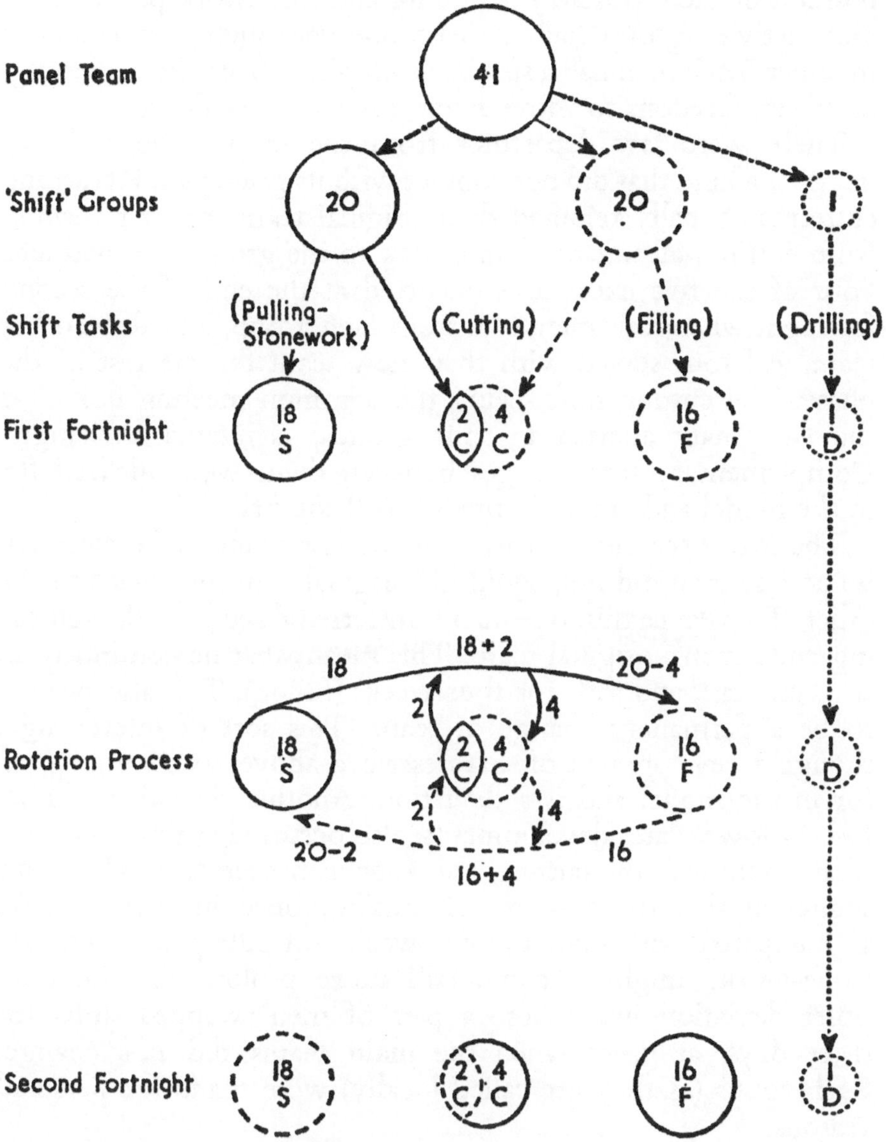

KEY
S Pulling and Stonework
C Cutting
F Filling
D Drilling
The figures indicate the number of men doing each task

position of men so that, for pulling and stonework particularly, each activity group had at least one thoroughly experienced member. Each man had a special work place; from the beginning, however, freedom to move from this was in evidence.

There were few departures from the model, the majority being of a kind that did not conflict with its principles. Returning cuttermen usually rejoined their original team: of 17 returning from cutting, all but five came back to the group they had left. Four of the five exceptions occurred at the end of the second fortnight when four men, two from each group, joined the other team. All four stayed with their new team for the rest of the phase. The cutting shift, being the common meeting point for the two main alternating shift groups, facilitated exchanges. Complementary interchanges between them were allowed for in the model and are indicative of its flexibility.

The fifth exception occurred at the last change of cuttermen, when one man did not rejoin his original team but went to the other. To balance this, one man transferred directly to the returning cutterman's original team. This meant that he continued on the same task and shift for four weeks (filling). This also proved to be a permanent change of team. This sort of interchange, though a development of that described above, was not allowed for in the model and is a deviation. Another deviation, which foreshadowed later developments, also occurred at this time: one man continued on cutting for four consecutive weeks. This suggested that the rotation of cutting, once in seven weeks as compared with once in two weeks for filling and pulling-stone-work, might become still more prolonged. The only other deviation was where a pair of men swapped shifts for three days, at times when the main teams did not change. Such coups (as they are called locally) were made for personal reasons.

That there were so few interchanges between the main groups, that four out of five were made in the first month, and that there were only three deviations, indicates that the team started in an almost fully-fledged state. It had been built up from two existing teams and centred on two captains who had taken part in the management-lodge discussions. They had the detailed knowledge on which to build a task-shift rotation system that embodied the intentions of these discussions, while at the same time permitting a degree of flexibility to meet personal preferences.

It was not until the ninth week that No. 2 Panel had to take account of the fortnightly changes in shift times. A change in the rotation system occurred in consequence (*Figure 10*) for similarity between periods 2 and 3 is low, but thereafter—until the end of period 9—successive periods resemble each other, phase 2 covering 28 weeks. Had the team made its fortnightly changes between weeks 10 and 11, 12 and 13, 14 and 15, etc., to coincide with the changes in shift times (*Figure 12* (*a*)), one group would have worked only fore and nightshift and the other only back. By staying, however, on the same task in week 9 as in week 8, and by changing tasks between weeks 9 and 10, 11 and 12, 13 and 14 etc., on the weekend between the fortnightly shift time changes, they shared all shifts in the ratio of two backshifts to one fore and one night (*Figure 12* (*b*)). This adjustment was established by week 13 and continued without change throughout the life of the panel. It is a development away from having task and shift changes every fortnight in the interests of an equal sharing of the good and bad shifts. The management-lodge discussions had not foreseen the consequence of alternating shift times when two panels started working.

The introduction of alternating shift times also gave rise to the need for a weekend cut when filling times changed from back to foreshift. If cuttermen changed every fortnight, the foreshift set would work 11 shifts and nightshift only nine. To overcome this, the team altered the task-changing period for cuttermen to four weeks. Next (period 4), they further excluded them from the rotation pattern, the roles being taken by men who tended to specialize. Three of the cutting roles were occupied by the same men during the 24 weeks of periods 4 to 9. The three other roles were occupied by six different men for two weeks each during period 4, and thereafter by only five different men, of whom two continued from period 4. There were thus only five changes in period 4 and only three in periods 5 to 9. If the four-weekly rota had been implemented there would have been 36 changes, whereas there were only eight.

With the change in the rotation of cutting, another way had to be found to balance the discrepancy between the 16 places on the

FIGURE 12
ALTERNATION OF SHIFT TIMES

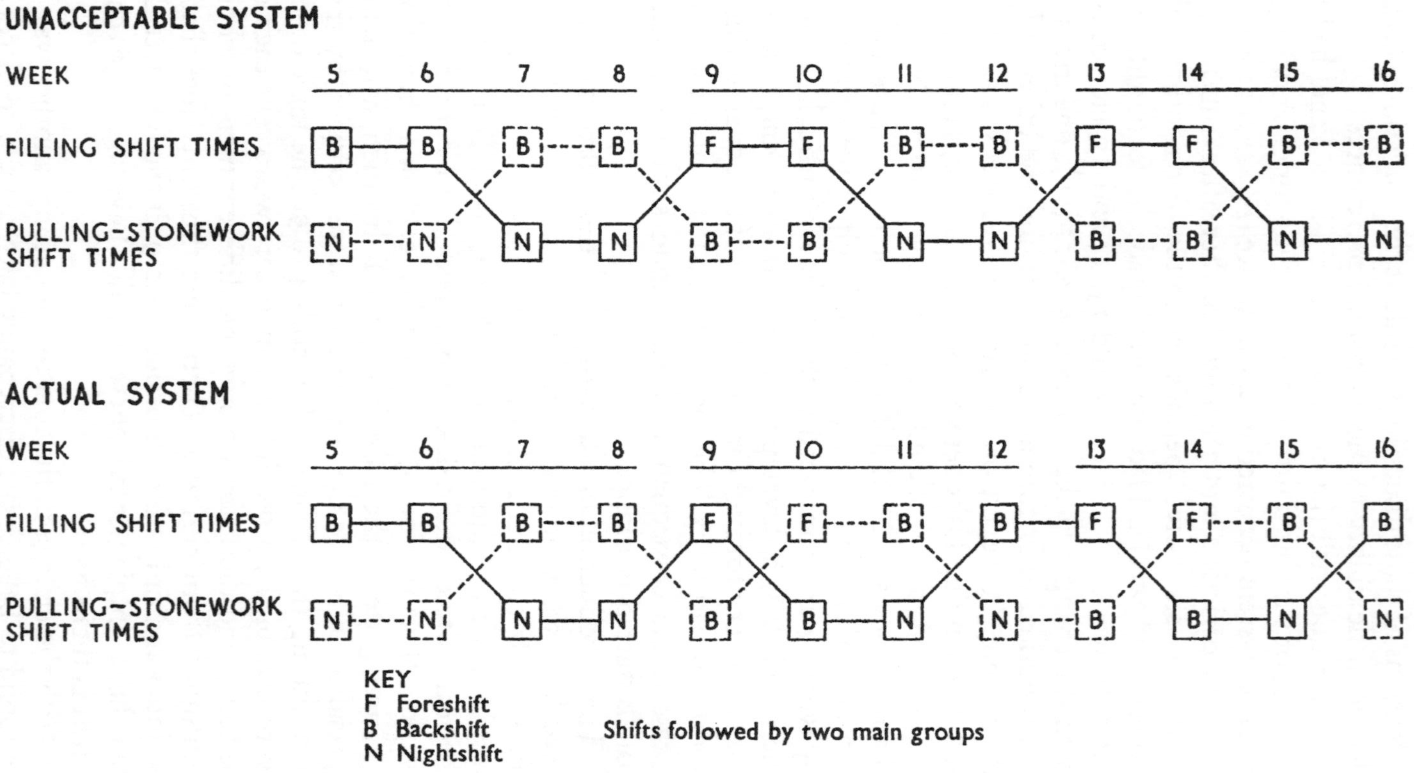

filling and the 18 on the pulling-stonework shift. The means adopted was that the group coming off pulling-stonework would leave two of their number to continue with the other group. On the first four occasions random pairs stayed behind, but from week 18 one man never moved and from week 32 he was joined by another man who also stayed on stonework. Between them these men account for half the 'stay-behind' vacancies and all such vacancies in later phases. This developing exclusion of two stonemen is a further example of the increasing segregation of small groups from the two main shift-changing groups. These changed arrangements were considered by the team as deliberate modifications of their system.

There were eleven occasions on which two men swapped shifts and seven when one man moved from his own into an absentee place on a preferred shift. These were sanctioned practices. Their occurrence indicates the extent to which the formal rotation model reflected the wishes of team members. There was only one instance of a true deviation, when in week 20 the panel ran with an extra man on filling and one short on pulling-stonework.

There were not many changes of task within shifts. In general, men kept to the same main task throughout. Only four, for example, split their time evenly between pulling and stonework. Distinct from this, of course, is the practice of task continuity (*vide* Chapter IX).

To sum up, in phase 2, No. 2 Panel modified its task-shift rotation system to ensure an equality of shift-sharing which would not otherwise have been possible with the change from fixed to alternating task times. The period of rotation for cutter-men was lengthened to four weeks but changes were rarely implemented. These modifications gave rise to quasi-permanent cuttermen and two quasi-permanent stonework roles to meet the discrepancy between the size of the filling and pulling-stonework groups. The increasing segregation of the driller, cuttermen, and two stonemen resulted in the creation of a strong gestalt *of two main groups of equal size who rotated filling and pulling-stonework.*[1]

[1] This type of development is in accord with 'structural balance' theory (cf. Heider, Cartwright & Haramy, op. cit.).

TRADITIONAL INFORMAL EXCHANGE (NO. 1 PANEL, EAST FACE, PHASE 1)

A few yards to the west of No. 1 Panel mothergate there was a fault running in a north-westerly direction, which cut off the coal. In order to open up an 80-yd face to the west, coal had to be extracted between the line of the mothergate and the fault for a distance of 90 yds. This was done by a 15-man team organized as a composite shortwall. When this preliminary work was finished, the West face was 90 yds ahead of the East face, which had not advanced at all. Whereas, therefore, No. 2 Panel started as a double unit, No. 1 Panel had to begin as a single unit. It worked as such for 13 weeks before becoming double—when the East face caught up with the West. Its rotation system came in consequence to be built up on the basis of two separate face groups, there being no possibility at the start of a panel-wide organization.

Though 41 men were allotted as the full team for the panel, only 23 were required until the East face caught up. These were cavilled from the 41, and Phase 1 covers the first 13 weeks when the panel was a single unit. During its first fortnight it filled on the foreshift. It then alternated back and foreshift filling every two weeks with No. 2 Panel.

When there is a change from back to foreshift filling an extra cut has to be made over the weekend. The East face team kept its work far enough ahead for this to be made by the stonemen on the Friday nightshift. From the beginning, therefore, the team was faced with the problem of developing a task-shift rotation system that fitted in with the alternation of shift times. The system they developed allowed weekly changes of job and shift but had no set pattern for groups or individuals. The men on the Friday foreshift, as they went out, would meet the Friday backshift coming in and settle the allocation of tasks for the following week. The men on the Friday nightshift were not at these meetings, but would make their preferences known through someone present. When cutting was on the nightshift, there would be 20 present and 3 absent. When stonework was on the nightshift there would be 12 present and 11 absent. What allocation a man obtained had not a little to do with his ability to talk others into letting him have it. Whether a nightshift man secured what he wanted in face of competition depended on the amount of deference others would grant him in his absence. When competing demands could not be reconciled, cavils would be 'rubbed', usually to decide who should take the undesirable shifts. Two principal factors determined the allocations: the effort of every man to get his own preference; the consideration that one should not grab for himself too big a share of the more desirable tasks or shifts.

This system bears no resemblance to that of the adjacent No. 2 Panel, which had already been working for six weeks. The reasons lie in the different origin of the teams. Cavilling the 41 men cut across the structure of existing groupings and, when the face commenced, the 23 men were an assembly of individuals without established leaders. Although a few may have worked together recently in different places, they had not worked together as a larger team. Moreover, no member had had direct contact with the negotiating group and no one else had told them of the management-lodge intentions.

In these circumstances they fashioned their system directly on their experience of composite shortwalls. A single face of 80 yds and a team of 23 men were not so different from a composite shortwall as obviously to call for a completely different method of distributing shifts, and a weekly 'face-to-face' meeting to sort out allocations for the following week was a shortwall practice. Whereas, however, on shortwalls the absent group was always a third of the team (5 out of 15), on single-unit longwalls it varies: at one time it was roughly an eighth (3 out of 23) and at another almost a half (11 out of 23). This, together with the rather larger team and the extension of the cycle over three shifts, created conditions too complex to be handled by continuing shortwall tradition.

During the first fortnight (weeks 7 and 8) every man did in week 8 the same task on the same shift as in week 7. Thereafter unplanned changes took place. Reference to *Figure 10* will show that during periods 4, 5, and 6, the percentage similarity between shifts in corresponding weeks was 63, 53, 61; for tasks it was 55, 46, and 48. Roughly 26 per cent of these similarities could arise by chance, leaving 20–37 per cent to be accounted for by the success of the men in getting their preferred shift or task in competition with others.

With a fixed relationship between tasks and shifts, it is not possible to know directly whether first preference was for task or shift. A number of pointers suggest that shift preference took precedence. In periods 4, 5, and 6 percentage similarity was always greater for shifts. In order to be on a shift he liked or to avoid a shift he disliked, a man would change his task. A comparison of the left- with the right-hand side of *Table 16* shows that the first man, for example, managed to avoid the foreshift completely; this, however, was not due to dislike of filling or cutting, for he spent half his time on these tasks when they were on other shifts. With shift times arranged as they were at this colliery, where the backshift of 9.30 a.m.–5 p.m. gives the equivalent of 'office hours', this shift was immensely more attractive than the other two, and the East face men made as many backshift openings for themselves as possible. *Table 15* shows that when filling was on the backshift more men were present so that there were fewer to do stonework on the nightshift. These anomalies resulted from the manoeuvres of a small clique who found themselves in a position to take advantage.

TABLE 15 VARIATION IN THE NUMBER OF MEN EMPLOYED DURING PHASE I OF THE EAST FACE ON FILLING AND STONEWORK SHIFTS

No. of men employed	*Number of Shifts* *Filling*		*Stonework*	
	Fore	Back	Night	Back
10	1	—	—	—
9	1	17	—	—
8	27	12	—	—
7	—	—	13	26
6	—	—	14	3
5	—	—	2	—
Total Shifts	29	29	29	29

Table 16 shows that the tendency to avoid a disliked shift was as important as that of seeking a preferred shift. The left-hand side lists for every man his proportion of fore, back, and nightshifts.

The lowest block records the shifts worked by the seven men who were the least successful in the weekly competition. Not only did they get a smaller share of the generally preferred backshift but also less chance to avoid whichever of the other two shifts they disliked most. By contrast, the seven men in the two top blocks managed to get a rather bigger share of the backshift and were also more successful in avoiding their most disliked shift (the fore

TABLE 16 PERCENTAGE OF SHIFTS AND TASKS WORKED BY MEN DURING PHASE I OF THE EAST FACE

	Shifts			Tasks				
Man	*Fore*	*Back*	*Night*	*Cutting*	*Filling*	*Pulling*	*Stone-work*	*Drilling*
A	—	53	47	24	24	15	37	—
B	2	56	42	20	24	56	—	—
C	8	60	32	24	38	8	30	—
D	8	53	39	6	23	—	71	—
E	8	38	54	9	17	—	74	—
F	52	48	—	—	—	—	—	100
G	33	58	9	16	35	2	47	—
H	15	60	25	—	35	17	48	—
I	16	54	30	8	39	29	24	—
J	18	52	30	16	18	18	48	—
K	19	58	23	17	23	19	41	—
L	26	56	18	9	42	—	49	—
M	38	44	18	10	52	24	14	—
N	29	47	24	15	23	38	24	—
O	28	44	28	47	19	18	16	—
P	33	43	24	—	58	33	9	—
Q	36	39	25	—	59	33	8	—
R	37	39	24	8	60	—	32	—
S	38	38	24	—	60	40	—	—
T	30	38	32	24	44	16	16	—
U	32	38	30	16	40	24	20	—
V	44	32	24	—	73	21	6	—
W	32	32	36	24	41	16	19	—
Theoretical Equality	25	44	31	13	35	17	31	4

in the case of the top five and the night in the next two). The remaining nine men worked a combination of shifts that was closer to theoretical equality, though five spent more than half their time on backshift.

By not having a fixed plan the East face team gained flexibility. They could not, however, ensure equality. Some men spent as much as 60 per cent of their time on backshift, others as little as 32 per cent. Time on nightshift varied between 0 and 54 per cent and on foreshift between 0 and 52 per cent. Equal shares would

TABLE 17 PERCENTAGE OF MEN CHANGING TASK AND SHIFT EACH WEEK DURING PHASE I OF NO. I PANEL, EAST FACE

	Type of Change			
Week	*Task and Shift*	*Shift only*	*No Change*	*Task only*
3	41	59	—	—
4	87	—	13	—
5	38	62	—	—
6	84	—	16	—
7	35	59	—	6
8	89	—	11	—
9	48	48	—	4
10	90	—	10	—
11	57	43	—	—
12	86	—	14	—
13	38	62	—	—
Average	43 87	55 —	— 13	2 —

have meant 44 per cent back, 31 per cent night, and 25 per cent fore. The weekly competition for preferred shifts was too vulnerable to manipulation to ensure equality. Nevertheless, only three 'coups' took place, all for one shift only, and there were only two instances of someone taking over an absent man's place on a different shift—the preferred backshift. These deviations, if they can be called such, were rare events, emphasizing the flexibility of the system in allowing a man to get what he wanted if he could convince the others.

From the point of view of the man in the team this was an unpredictable system. He did not know what he was going to be doing the following week. Yet, looked at as a whole, it had a certain regularity—enough to get the cycle completed.

Table 17 shows the amount and kind of change that occurred from week to week. A regular pattern runs through the phase. This regularity does not arise because any men or group were following a fixed plan. The individuals involved on each occasion were usually different. Moreover, they were unaware that they were contributing to a regular pattern. Each man was aware only of making his own arrangements.

Lacking any direct contact with the negotiating group or any explicit direction regarding task-shift rotation, the East face developed a system which had its origins in composite shortwalls. In face-to-face meetings between the Friday fore and backshifts, at which an absent nightshift of varying size was only indirectly represented, the men competed with each other for their preferred shifts. Though the system as a whole was flexible and got the work done, its weakness lay in its inability to ensure equality. It departed from the principle of a fortnightly change of task and shift, no provision having been made to ensure that this principle would be followed or even that the outcome of the management-lodge discussions was known.

CHAPTER XVII

The Appearance of Independent Developments[1]

PARTITIONING (NO. 1 PANEL, WEST FACE, PHASE 1)

In week 20, when the East had reached the line of the West face, No. 1 Panel was ready to start as a double unit. Of the 18 West face men who had been working elsewhere as a composite team, one became the driller for the panel as a whole and was attached to the East face while three men from the East joined the remainder to form a 20-man team for the West. From the start, the West had a planned and stable system of task and shift rotation. Of the 20 team members, 16 had worked as substitutes on No. 2 Panel and knew something of its system. A certain number had also worked as substitutes on the East face, which had shown them the limitations of attempting to carry on in the old way. Having all worked together on a shortwall, they had had ample opportunity to discuss among themselves what system they should develop. This was finalized at a meeting held in the colliery office before the face started.

Though influenced by No. 2 Panel, differences arose because they were a face- and not a panel-size group, and because of the particular circumstances in which they had gained knowledge of No. 2 Panel's system—by acting as substitutes. The result was a pattern very much their own (*Figure 13*). Its characteristics may be summarized as follows:

(a) No. 2 Panel was beginning to treat the six cuttermen as quasi-permanent, one task—two shift men. The three West face cuttermen were similarly treated.

(b) No. 2 Panel required two men to stay on pulling-stonework when their group moved to filling. On the West face, two

[1] Source paper: 7.

FIGURE 13

ROTATION SYSTEM OF NO. 1 PANEL, WEST FACE, PHASE 1

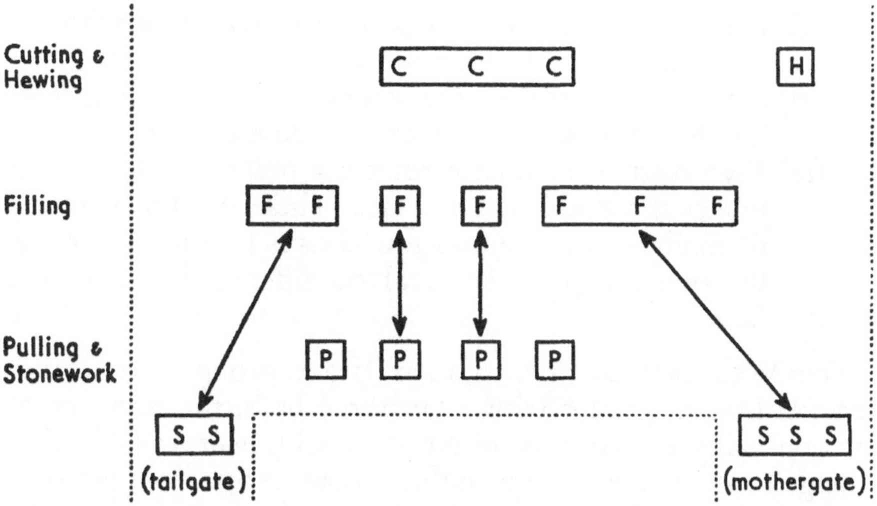

KEY
Each letter represents a man and a task
↔ Connects men or groups who rotate the tasks shown on three shifts
Other men work at one task on two shifts

pullers decided to stay permanently on pulling—they wished to avoid the foreshift. These two, however, did not work together; each worked with two other men who alternated pulling and filling.

(c) No. 2 Panel created two 16-man panel-wide shift groups who rotated tasks and shifts. The West face was left with two unequal groups—eight fillers and seven stonemen. To equalize these, one filler took on the quasi-permanent role of mothergate hewer. He worked the same shifts as the cuttermen, even though the East face hewer worked on the filling shift. This isolation was a personal preference.

(d) The No. 2 Panel principle of two main panel-wide groups rotating tasks and shifts would scarcely be evident to men working as substitutes at infrequent intervals, on different shifts, and at different tasks. Substitutes tended to see the system as composed of small activity groups such as mothergate stonemen or tailgate stonemen. Features (a)–(c) are consistent with this type of perception, and West face men conceived their system on the principle of territorially-based activity groups:

(i) Four men rotated two tailgate stonework jobs with the two filling positions nearest the tailgate.
(ii) Six men rotated three fore caunch stonework tasks with the three filling positions nearest the mothergate.
(iii) Two pairs of men each rotated a pulling with a filling job in the middle of the face. These did not perceive themselves as a group of four, but as two pairs, because the two pulling tasks involved different halves of the face.

The West face partitioned their larger group into smaller distinct rotation groups with a territorial basis, the sequence of rotation being the same as on No. 2 Panel (*Figure 12 (b)*). This arrangement, while giving rather more scope for personal preference than No. 2, ensured 'fair shares' by eliminating the 'competition' characteristic of the East face. The introduction of separate sub-groups as the basis of task- and shift-sharing made the pattern of change specific for each individual.

Phase 1 of the West face (periods 6, 7, and 8) lasted 13 weeks and achieved a high level of consistency (*Figure 10*). There was, however, flexibility through swaps and movements into the places of idle[1] men on preferred shifts. Swaps were more usual between men in different activity groups, showing that they regarded themselves as members of a total team rather than simply as members of their rotation sub-group. The degree to which tasks done corresponded with those laid down is shown in *Figure 14*. In the first week there was 74 per cent conformity; thereafter, it remained above 80 per cent, except for week 31. This is high when allowance is made for the absence during the whole phase of one injured man. His absence as such does not affect percentage conformity, but offers other men the temptation to take his idle place on the backshift, which reduces percentage conformity by 5 per cent. The lower conformity in week 31 illustrates the flexibility of the system; of the 27 per cent nonconformity in that week, 20 per cent is due to swaps involving two men, and 7 per cent to a man moving into the backshift place of an absentee.

[1] In local usage 'idle' is an exact synonym for 'absent'.

The greater part of the West face team was working together as a shortwall composite group just before going on to the longwall face. They had also worked as substitutes on No. 2 Panel, whose influence is apparent in the rotation system they devised. Differences arise, however, due to their being a face- and not a panel-size group, and to the limited opportunity they had to grasp the basic features of No. 2 Panel. Their own system is, therefore, distinctive and constitutes an innovation. It is based on the partition of the total group into a number of smaller, separate, territorially-based activity groups.

FIGURE 14

PERCENTAGE OF SHIFTS WORKED IN WHICH THE TASK ALLOCATION CONFORMED TO ROTATION SYSTEM MODEL

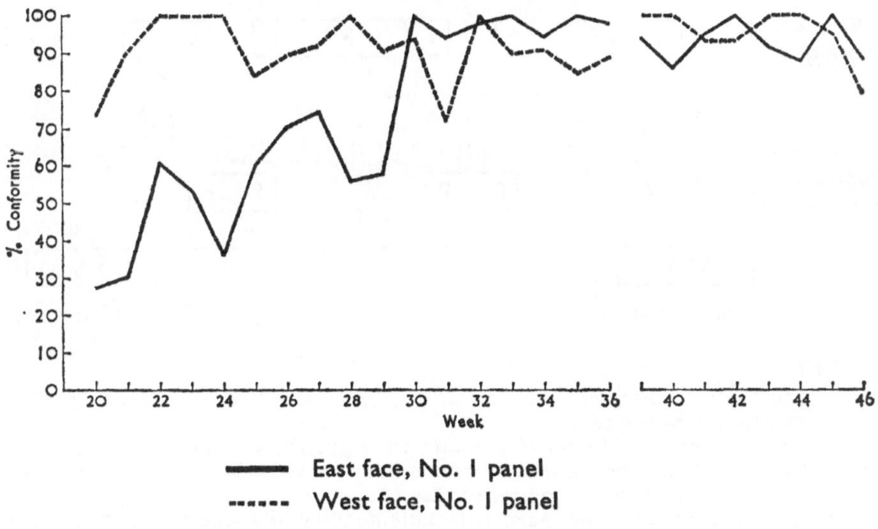

COMPLEMENTARY SHIFT-SHARING (NO. I PANEL, EAST FACE, PHASE 2)

When No. 1 Panel became a double unit, the East face had its numbers reduced from 23 to 21, the driller remaining with them. Period 6, during which the double unit came into operation, resembles the earlier periods. Period 7 shows a definite change, indicating that work was organized in a different way. Period 8 resembles period 7 to a degree greater than any two previous periods and thereafter the resemblance between successive periods stabilizes at a very high level. A process of change was at work during periods 6, 7, and 8. The explanation was supplied by the men themselves. At the time the panel became a double unit they were becoming dissatisfied with the way their competitive system was leading to inequality in the distribution of the more attractive shifts, and the example of guaranteed equality they

FIGURE 15

ROTATION SYSTEM OF NO. 1 PANEL, EAST FACE, PHASE 2

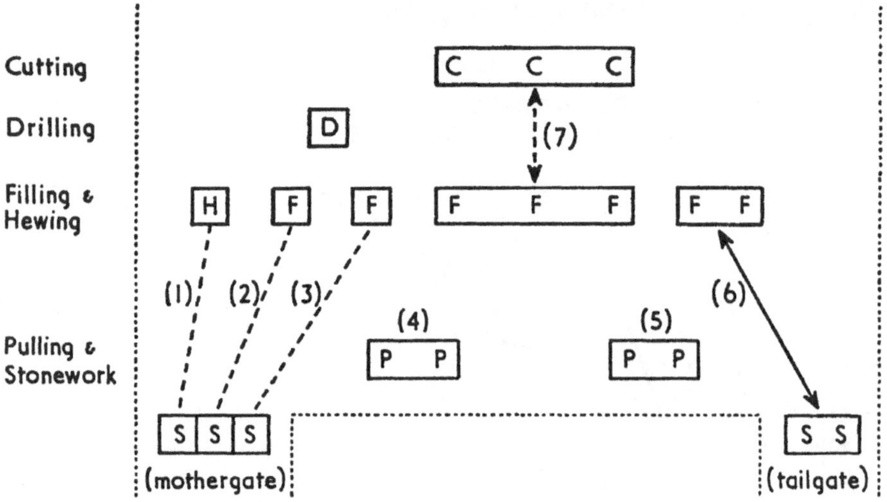

KEY

Each letter represents a man and a task; the figures refer to groups mentioned in the text

◄──► Connects groups of men who rotate the tasks shown on three shifts

------ Connects men who have a complementary shift-sharing arrangement; each man works at one task on two shifts

◄--► Connects groups who have a complementary shift-sharing arrangement with two tasks (see text)

became acquainted with through the West face brought their dissatisfaction to a head. When changes began to take shape it was the West face which provided the model (*Figure 15*).

As seen by East face men, partitioning was the most obvious feature of the West face but the way in which they divided up their tasks developed out of discussions which took place between the various small groups who worked together rather than as a result of a decision by the team as a whole. Six of the seven sub-groups formed used a method of sharing tasks and shifts which differed radically from that of the West face. Their competitive system having failed to guarantee fair shares to all, they developed in complementary shift-sharing a means of meeting this deficiency:

(a) Group 1 comprised two men, one of whom disliked the nightshift and the other the foreshift. Taking as their tasks mothergate hewing and a back caunch stone place, the first stuck to hewing fore and back, the second to stonework back and night. Groups 2 and 3 worked the same way with a filling task in place of hewing. The six men in Groups 1, 2, and 3 all followed the same pattern and could be considered as one six-man group, but they saw themselves as three pairs, very much in the 'marrow' tradition.

(b) Groups 4 and 5 were each made up of pairs of pullers who worked together. They stayed permanently on pulling following the fortnightly shift changes, doing equal amounts of back and night, and avoiding foreshift.

(c) Group 6 was a four-man group sharing the two tailgate stonework tasks and the two neuk-end filling positions. They shared all shifts equally, getting two back to one fore and one night, and were organized in the same way as the corresponding West face tailgate group.

(d) Group 7 used the principle of complementary shift-sharing but a little differently from Groups 1 to 5. Three of the men worked only foreshift, doing cutting one fortnight and filling the next. The other three changed from filling on back to cutting on night. Each did equal amounts of cutting and filling, three men only on fore, the other three alternately on back and night. Group 7 was distinctive, being transitional between Groups 1, 2, 3 and Group 6. It had equal task-sharing like Group 6 and complementary shift-sharing like Groups 1, 2, and 3.

Figure 14 illustrates the growth process of this revised system in more detail than can be inferred from *Figure 10*. It was not until week 30 that substantial conformity was reached which was maintained throughout the remainder of phase 2. Between weeks 22 and 30 there was a period of fluctuation accompanied by increasing conformity, which corresponds to the spread of the revised system across the face. The records show that the men adjacent to the West face—Groups 1, 2, and 3 in the mothergate—were the first to change. These were followed by the two pairs of pullers (Groups 4 and 5) and then by the group farthest from the mothergate (Group 6). The last group to emerge was Group 7, which combined three cutting roles with the three filling roles in the middle of the face 'left over' from the other groups. The distinctive character of its sharing system suggests that some time was required to arrive at a method that would give each man his preferred combination of shifts and tasks. In each of these four 'stages', (a)–(d), a different scheme of sharing tasks and shifts was used, indicating that the solutions were invented locally by the emerging groups, there being no overall face 'policy' formulated by the group as a whole. The East face took nine weeks to achieve its ultimate system, whereas the West took only three. The East took this much longer to 'unwind' from the old and develop a new way.

This rotation system endured without change for the remainder of the life of the panel, the few changes which did occur being merely changes in the people who occupied certain roles. There were three such occasions—in periods 12, 15, and 18—which account for the slight drops in similarity between periods (*Figure 10*). The first involved four roles, the second and third, two. Only six men were concerned, one of them on all three occasions, which suggests that the teams were sufficiently satisfied with their roles (and with the opportunities to enrich them through the occasional coup or by taking the place of absentees) not to wish further change. Throughout the settled part of the second phase (period 9 onwards) the team worked very closely to the model, though following period 16 the level of conformity fell steadily. This was due largely to a reallocation of men to meet disturbances caused by bad roof conditions associated with lost cuts. The generally high level of conformity, the absence of changes in the rotation system model, and the infrequency of role changes suggests a rigidity in phase 2—a swing of the pendulum away from the 'anarchy' of the previous phase.

Awareness of the West face, which, while permitting personal preferences, ensured fair sharing of good and bad shifts, brought to a head increasing dissatisfaction among East face men with their original system. The West face provided a model which they adapted by means of complementary shift-sharing to enable men to avoid the shift they disliked most. The change spread slowly from the mothergate across the face, each territorially-based activity group devising a method of sharing shifts and tasks to suit its members. The team conformed to a very high degree to this model—which remained unchanged during the rest of the life of the panel—as if to protect themselves from the gross inequalities which resulted from their phase 1 attempt to continue traditional short-wall informality.

CHAPTER XVIII

Convergence through Mutual Learning[1]

NO. 2 PANEL (SHIFT EXCHANGE), PHASES 3 AND 4

After the annual pit holiday which followed period 9, No. 2 Panel further modified its rotation system (cf. the much lower level of similarity between periods 9 and 10). Our informants said that the natural break afforded by the holiday enabled the team to reassess the work situation. The faces were farther in, the standard of maintenance by seam personnel on more worn equipment had been reduced, and interference had increased. Accordingly they became more vigorous in their attitude towards claiming for waiting-on and shift work.

They began to re-implement their policy of four-weekly cutting changes, with a complete change of cuttermen in period 10. This came about because the men who had done the cutting during phase 2 had become 'fed up' with not getting any back-shift. The two main shift groups differed, however, in their method of rotating these roles. Group (a) formally divided itself into sets of three, one set going on to cutting every four weeks. In group (b) one man became a quasi-permanent cutter (in the 75 weeks of the panel's life he spent only six weeks away from cutting), while two additional men were cavilled every four weeks from those who had not been on cutting to make up a set of three. One group had a planned, the other a traditional, approach to the establishment of equality. At this time the two faces were also differently equipped, the East with timber props and the West with adjustable steel props, which were heavier to handle. When filling, group (a) changed faces every week, moving from the 'steel' to the 'timber' face to share the heavier work, whereas in group (b) men spent both filling weeks on the same face. At first sight it may seem surprising that the two

[1] Source paper: 7.

main shift groups of a team which had organized itself on a panel-wide basis should differ in the extent to which they felt the need for formal rotas to ensure equality. As, however, they were based on two previously existing shortwall groups, men of similar outlook had marrowed themselves together round two rather different team captains, and the groups differed in their attitudes. When both faces were equipped with steel props, the changeover still continued in group (a). It was not carried out, however, during pulling and stonework because, more than on any other shift, continuous knowledge of a particular gate, caunch, or face is important for control of the roof.

The beginning of a fourth phase is indicated by a rise in similarity between periods 13 and 14 to a level between phases 2 and 3. With the exception of a marked fall in period 17—which did not result from a system change—this level was maintained. As during phase 2, though more gradually, the changes in the cutting rotation became less frequent, more men tending to spend longer than four weeks. Group (a) reintroduced a quasi-permanent cutterman (period 14) and thus resembled group (b). Later (period 17) both groups introduced a second quasi-permanent man, and by the end of the life of the panel the indication was that the third man would also cease to change each month. The explanation of these changes may be sought in their timing. The one major technological change, that of increasing the length of the cutting jib, was introduced in week 60 (period 15). The longer jib, which the team knew about, was also preceded by a floor roll which made cutting more difficult. Under such conditions at least one experienced cutterman is required. The group took action accordingly, and added a second quasi-permanent cutterman when geological conditions became still worse and cuts were lost.

In period 14 the first instance occurred on No. 2 Panel of complementary shift-sharing. The man responsible disliked the foreshift and had previously succeeded in swapping foreshift filling for nightshift cutting. In period 14, he arranged with a man from the other shift group to remain on stonework night and back, while his partner stayed on filling back and fore. This arrangement was a personal one and both men went on to cutting at different times as required by the rota. On these occasions the man not on cutting made a complementary arrangement with someone else, by couping. It was not until period 17 that another pair of men made a similar arrangement, sharing filling and stonework in the same way. This extension of complementary shift-sharing accounts for the drop in similarity between periods 16 and 17. The idea was learned from No. 1 Panel where it had now become widespread on both faces.

The non-implementation of the cutting rotation is the first instance of a panel team modifying its system to meet the changing demands of the technology and the work environment. The introduction of complementary shift-sharing by two pairs is yet another instance of the spread of rotation practices—but in this case from one panel to another. The idea was not taken up by the panel as a whole, but was sanctioned as a personal arrangement which did not conflict with their basic principles. With the quasi-permanent driller, cuttermen, and stonemen and the two pairs of men with complementary shift-sharing arrangements, a marked convergence had taken place in the rotation systems of the two panels.

NO. 1 PANEL, WEST FACE (PARTITIONING), PHASES 2, 3, AND 4

In week 33, shortly after complementary shift-sharing was established on the East face, two of the groups on the West introduced it, marking a second phase in their development. Of the six men in the mothergate stonework and filling group, three stayed permanently on stonework (back and night) and three permanently on filling (fore and back). The pulling-filling pair nearest the tailgate made similar arrangements, one man staying permanently on pulling while the other stuck to filling (*Figure 16*). In much the same way as the East face gradually took over partitioning, the West now took over complementary shift-sharing, and the two faces became more alike.

In week 35, the quasi-permanent driller, hitherto from the East face, went on to filling in the place of the permanent West face absentee and for the first ten weeks of the phase the drilling rotated freely over the whole panel. This showed that the men on both faces saw themselves not only as East or West face men, but as members of a 41-man panel team with a single shared job. There were also a few changes in the occupants of particular roles. The first occasion (period 12) concerned a cutting and a

FIGURE 16

ROTATION SYSTEM OF NO. 1 PANEL, WEST FACE, PHASE 2

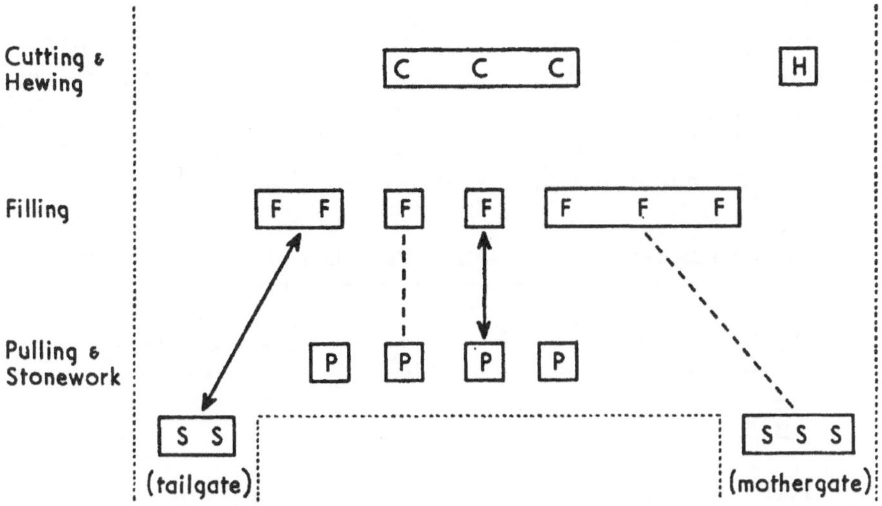

KEY

Each letter represents a man and a task

←→ Connects men or groups who rotate the tasks shown on three shifts

------ Connects men or groups who have a complementary shift-sharing arrangement; they and the other men work at one task on two shifts

filling-stonework role. The second, brought about by the return of a long-term accident case, involved three men occupying quasi-permanent filling, cutting, and stonework-filling roles.

The last week of period 14 marked the beginning of a new phase, which preceded by three weeks the introduction of the longer cutting jib. The men knew of the intention to introduce the longer jib but did not make changes entirely for this reason, although this helped to speed them up. For some weeks the two men who had been permanent cutters and who had never worked backshift had been pressing for a change. An arrangement, similar to that on the East face, was made between the three cuttermen and three permanent fillers (*Figure 17*). The former took backshift filling and nightshift cutting, the latter the purely foreshift job of cutting and filling, so that this rotation group had equal task-sharing and complementary shift-sharing. Further convergence of the rotation systems of the two faces had thus occurred.

Towards the end of phase 2, a development took place which became clear from phase 3 onwards. Four men who combined tailgate stonework with filling rotated to get two backshifts to one fore and one night. Two of the men followed this pattern according to the rule, the other two in an *ad hoc* manner, yet they can be looked on as a complementary shift-sharing pair. As required, they manned one stonework and one filling role but in a way that suited their preferences (*Table 18*).

TABLE 18 DISTRIBUTION OF TASKS AND SHIFTS BETWEEN TWO MEN

	Shifts			*Tasks*	
Man	*Fore*	*Back*	*Night*	*Stonework*	*Filling*
A	1	18	6	19	6
B	13	7	5	6	19
Total	14	25	11	25	25

Man A does not like the foreshift and is keen on the back, whereas Man B does not mind *not* being on the back, nor does he mind being on the fore. They have complementary preferences and work this to their advantage, but without upsetting the whole rotation system.

The final modification of the West face took place during period 17. Though small, it was definite and brought the West more closely to resemble the East. The remaining rotation of a shared filling-pulling role was broken up to make the fourth pulling role permanent and to give another permanent filling role making five of the seven filling roles permanent (*Figure 17*). Throughout phase 4 and the preceding phase, the level of conformity of the West face team to their rotation model was high and above that for the East. Unlike the East, the West did not show so marked a decline in conformity during the last few periods, since absenteeism—due to accidents—was lower and there was consequently less opportunity for men to move into the places rendered vacant, usually on the preferred backshift.

In this final phase, the West face again followed the East in having four permanent pullers, thus further increasing their resemblance. Though it conformed more closely to its model than the East, it showed

FIGURE 17

ROTATION SYSTEM OF NO. I PANEL, WEST FACE, PHASES 3 & 4

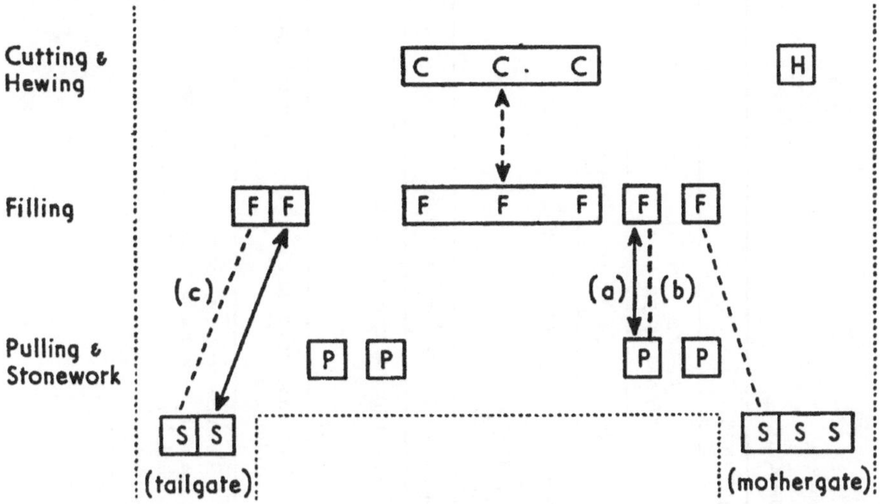

KEY

Each letter represents a man and a task

◄—► Connects men who rotate the tasks shown on three shifts

------ Connects men or groups who have a complementary shift-sharing arrangement; they and the other men work at one task on two shifts

◄--► Connects groups who have a complementary shift-sharing arrangement with two tasks

(a) Phase 3
(b) Phase 4
(c) Modified complementary shift-sharing

a greater facility in altering its system—having four phases compared with two. The West face team, having deliberately designed their rotation system from the outset, achieved desired change by altering the model, whereas the East face which developed under the stress of strong internal pressures, achieved it only by role exchange on a personal basis.

GENERAL OBSERVATIONS

Tables 19 and *20* summarize the main points in the development of the task-shift rotation systems of the two panels and facilitate comparison of the conditions which gave rise to the particular systems developed, bringing out their underlying principles and their characteristic features. In addition, they show the distribution of task-shift roles resulting from the rotation systems—the

TABLE 19 Rotation System Summary

	NO. 2 PANEL			
	Phase 1 (Weeks 1–8)	*Phase 2 (Weeks 9–38)*	*Phase 3 (Weeks 39–54)*	*Phase 4 (Weeks 55–77)*
ORIGIN OF TEAM	20 men from team of 23, 13 from 20, 4 from 15, 4 from 15 = 41 men			
IMMEDIATE PRE-LONGWALL EXPERIENCE	20 men from team working a 70-yd. composite shortwall 15 ,, ,, ,, ,, ,, 40-yd. ,, ,, 4 ,, ,, ,, ,, ,, 30-yd. ,, ,, 4 ,, ,, ,, ,, ,, 30-yd. ,, ,,			
QUALIFICATIONS	All qualified in all tasks, except 15 men not qualified in cutting			
CONTACT WITH NEGOTIATING GROUP	Direct, through 2 team captains participating in negotiations			
SOURCE OF ROTATION SYSTEM	Management-Lodge discussions.	Phase 1.	Own innovation on basis of Phase 2.	No. 1 Panel faces and own Phase 2.
METHOD OF GAINING KNOWLEDGE	Through two team captains who took part in discussions.	—	Re-assessment of situation in discussion after pit holiday.	Contact with No. 1 Panel team and own experience.

MODIFICATIONS MADE TO SOURCE SYSTEM	Quasi-permanent driller.	Tasks change in between shift time changes. 4-weekly rotation period for cuttermen, developing quasi-permanent role. 2 men quasi-permanently on stonework.	Re-implementation of 4-weekly cutting changes and introduction of cutting rota by one shift group, cavilling method by other. Rotation of faces for filling by one shift group.	Development of quasi-permanent cutting roles. Complementary shift-sharing of filling and stonework roles introduced by two pairs of men.
CHARACTERISTICS OF THE ROTATION SYSTEM	Equal task and shift-sharing between two panel-wide shift groups. Change of task and shift every fortnight. System specifies rules for rotation of main groups not individuals.	Equal shift-sharing in two main groups. Rotation to give 2 back shifts for 1 fore and 1 night shift. System specifies rules for rotation of groups. Men return to same work place on successive filling and stonework shifts.	Essentially as in Phase 2, with in addition equal sharing of cutting shifts. One shift group rotates faces on filling shift ensuring equality of working conditions.	Equal shift-sharing as in Phase 2; equality of working conditions on filling shift in one shift group. Complementary shift-sharing by four men not in conflict with basic rotation pattern.
TASK-SHIFT ROLES RESULTING FROM THE ROTATION SYSTEM (Driller excluded)	3 tasks, 3 shifts: 40 men	1 task, 2 shifts: 2 men 2 tasks, 3 shifts: 38 men LATER: 1 task, 2 shifts: 8 men 3 tasks, 3 shifts: 32 men	1 task, 2 shifts: 2 men 3 tasks, 3 shifts: 38 men	1 task, 2 shifts: 12 men 2 tasks, 3 shifts: 28 men

TABLE 20 ROTATION SYSTEM SUMMARY

	NO. 1 PANEL, EAST FACE		NO. 1 PANEL, WEST FACE			
	Phase 1 (Weeks 7–21)	*Phase 2 (Weeks 22–83)*	*Phase 1 (Weeks 20–32)*	*Phase 2 (Weeks 33–57)*	*Phase 3 (Weeks 58–67)*	*Phase 4 (Weeks 68–83)*
ORIGIN OF TEAM	Three 15-man teams less 4 men = 41 men					
	23 men cavilled from the 41.	20 men from Phase 1 team plus 1 man from West Face = 21 men.	Remaining 18 men from 41 men, less 1 man to East Face, plus 3 men from East Face Phase 1 team = 20 men.			
IMMEDIATE PRE-LONGWALL EXPERIENCE	23 men had not worked together as a team, but in three different groups on composite shortwalls.	—	18 men worked as a team in a composite shortwall.			
QUALIFICATIONS	All qualified in all tasks, except 7 men not qualified in cutting.	All qualified in all tasks, except 5 men not qualified in cutting.	All qualified in all tasks, except 11 men not qualified in cutting.			
CONTACT WITH NEGOTIATING GROUP	None		None			
SOURCE OF ROTATION SYSTEM	Composite shortwall practice in the seam.	West Face, Phase 1.	Own innovation on basis of No. 2 Panel, Phase 2.	East Face, Phase 2.	East Face, Phase 2 and innovation by two men.	Own Phase 1.
METHOD OF GAINING KNOWLEDGE	Experience common to all the men.	Contact between face teams. Spread across the face.	16 of 20 men served as substitutes on No. 2 Panel.	Contact between face teams.	Contact between face teams on cutting shift.	—

MODIFICATIONS MADE TO SOURCE SYSTEM	None, except quasi-permanent driller.	Complementary shift-sharing, increasing number of quasi-permanent jobs.	Partitioning of team into territorially based sub-groups. From start, quasi-permanent hewer, 3 cutters and 2 pullers.	Panel wide rotation of drilling on 1 to 2-weekly basis (in conjunction with East Face). Partial introduction of complementary shift-sharing.	Complementary shift and task sharing extended to cutters, reducing number of quasi-permanent jobs. *Ad hoc* complementary sharing operated by two men.	Application of complementary shift-sharing to remaining shared pulling-filling role.
CHARACTERISTICS OF THE ROTATION SYSTEM	*Ad hoc* changes of task and/or shift made at weekly meeting. All team not present at meeting. Balance attempted between personal preference and fair sharing of good and bad shifts. No explicit rules for individuals.	Small complementary shift-sharing groups to meet personal task and shift preferences and guarantee fair sharing within small groups. Specifies rules for individual changes. Fixed positions for work on filling shifts.	Small task rotation groups to allow personal preference for tasks and equal shift - sharing. Specifies rules for individual changes. Fixed positions for work on filling shifts.	Equal task and shift-sharing in some groups; complementary shift-sharing in two groups. Specifies rules for individuals. Fixed positions for work on filling shifts.	Equal task and shift - sharing between two pairs, complementary shift-sharing among the rest. Specifies individual rules for task and shift except between one pair in which short term preference operates. Fixed positions for work on filling shifts.	As in Phase 3, except that only one pair share task and shift equally.
TASK-SHIFT ROLES RESULTING FROM THE ROTATION SYSTEM (Driller excluded)	4 tasks, 3 shifts: 22 men	1 task, 2 shifts: 10 men 2 tasks, 1 shift: 3 men 2 tasks, 2 shifts: 3 men 2 tasks, 3 shifts: 4 men	1 task, 2 shifts: 6 men 2 tasks, 3 shifts: 14 men	1 task, 2 shifts: 14 men 2 tasks, 3 shifts: 6 men	1 task, 2 shifts: 8 men 2 tasks, 1 shift: 3 men 2 tasks, 2 shifts: 3 men 2 tasks, 3 shifts: 6 men	1 task, 2 shifts: 10 men 2 tasks, 1 shift: 3 men 2 tasks, 2 shifts: 3 men 2 tasks, 3 shifts: 4 men

combination of tasks and shifts that each man would work if the systems were followed.

No. 2 Panel and No. 1 Panel (West) worked to systems which they introduced from the start and which subsequent modifications did not radically alter. No. 1 Panel (East) developed two different systems, as their first phase attempt was unsatisfactory. This divergence in development is correlated with basic differences in the origin and structure of the teams. No. 2 Panel was made up of two large sub-groups who had worked on shortwalls as independent teams; in developing their rotation system they made use of this existing organization in the form of two panel-wide shift-alternating groups. No. 1 Panel (West) had also been working as a group whose composition remained virtually unchanged when it went on to the longwall face. No. 1 Panel (East) was an *ad hoc* group which had not been welded together.

The first rotation system to develop matched closely the original intention of the management-lodge discussions, but as the other faces opened up successive modifications were made which departed from this formulation. Management and lodge were aware of these changes in only the vaguest way and at no time did either consider it necessary to find out what had happened. The modifications which the teams introduced reflected their appreciation of the systems already working on the other faces. They adopted those features most readily perceived without appreciating the reasons which gave rise to them or the more general principles of their structure.

A major factor which allowed such varied developments to take place was the absence of any formal channel of communication between either the management or the lodge sides of the negotiating group and the men forming the teams. Only at the beginning of No. 2 Panel was there, for fortuitous reasons, such a link; thereafter teams developed their systems as they chose. The records kept by the deputies informed seam and pit management how the various roles were being manned. Since, however, work was being carried out according to the agreement and the panels were functioning smoothly, management was content that the teams should develop in their own way.

Information about the working of the rotation systems was

always available to the lodge through team members who served on the committee. When new agreements for other seams were under discussion, these men were called upon during lodge meetings to explain details of their methods of working. Yet at no time did they give a general account of their rotation systems. One consequence was that when composite longwall working was introduced into another seam, none of the newly constituted teams adopted task-shift rotation systems resembling those of the Manley panels. They were unaware of their nature and, apart from one long-established group who rotated shifts on the shortwall pattern, the only form of task-shift rotation to develop was on a limited personal basis (Chapter XXV).

No. 2 Panel, whose team captains had been in contact with the negotiating group, maintained the original composite conception of multi-task-multi-shift work roles and made maximum use of the experience of the men. No. 1 Panel, developing different systems, tended to restrict the variety of main tasks and shifts that a man might work. Nevertheless, these groups all worked in the composite tradition by practising task continuity. The development of partitioning and complementary shift-sharing suggests the possibility that men on No. 1 Panel might become more identified with their sub-groups than with the team as a whole and so lose the sense of cycle responsibility. This theme was discussed in Chapter XIV, which examined how the two panel teams regulated their activities when working conditions deteriorated. Its more restrictive organization subjected No. 1 Panel to greater strain, but commitment to the overall goal was maintained.

All three systems worked, in the sense that loss of cycles was extremely small, though available indicators consistently demonstrated some superiority in the performance of No. 2 Panel. Those who created the systems said in each case that they were satisfied with their own pattern and would not change to either of the others. That such a range and wealth of social creativeness should have appeared in comparable units which were physically adjacent is of general interest to a theory of the autonomous work group. Experience on these panels lends no support to the belief that a group of 41 men is too large to organize itself effectively or adapt to changes in the task environment, while the existence of three different and spontaneously developed means of successfully attaining the same end provides an illustration at social level of the principle of equi-finality in open systems (cf. von Bertalanffy, op. cit.).

SECTION SIX

The Interaction of Management, Trade Union Lodge, and Working Group in a New Situation

---

CHAPTERS

CHAPTER XIX

Initial Failure[1]

THE CHARACTER AND BACKGROUND OF THE PROJECT

This Section describes the course of events in the development of a double-unit hewing longwall organized on composite principles. In its early life the geological conditions were difficult in the extreme, and in the team, which was a new group put together for extraneous reasons, the majority of the members had no previous experience of either hewing or longwall faces. Each set of circumstances, geological and socio-psychological, aggravated the other. Conditions could scarcely have been more different from those obtaining on the cutting faces described in Section Five, which were in the same pit.

The panel comprised the first two faces of a new drift leading in from the surface, so that work began near the outcrop. This was a major factor in creating difficult mining conditions, which improved when the faces got further in and cover became more substantial. The haulage was simple: sets of tubs went by endless rope straight from the loading-point to the screening plant, some 250 yds, and returned immediately. From the screening plant the coal was transported to the washery by lorries. The drift was separate from the other workings of the colliery and the double unit was a new enterprise expected to produce 1,000 tons of coal per week. This was equal to 25 per cent of the previous output of the whole colliery, whose life it would considerably extend.

The panel consisted of two 80-yd faces, one on each side of a mothergate. The line of advance was northwards, parallel with the cleat of the coal, and the East face was in advance of the West by 4′ 6″. Both top and bottom caunch were taken, with the fore caunch ahead of the leading face by 9′ 0″. The coal was 28″ in section, with no band, but with a 4″–8″ layer of ramble above it.

[1] Source paper: 17.

Bottom loading face belts fed on to a gate conveyor, which discharged into a hopper from which the tubs were filled. Face supports were wooden props and steel straps, with collapsible steel chocks—the innovation which had solved the problem of roof control in the Manley. The coal was won by pneumatic picks supplied with compressed air from a pipe running along the face in front of the conveyor.

A complement of 51 faceworkers was planned, with six 'spare' men to provide substitutes in case of absences. There was also a six-man team on development work.

The emergence of composite working at this colliery was discussed in Chapter VIII. Success in the Manley led to a decision to run the new drift on similar lines, modified for a hewing technology. Both management and men expected 'teething troubles', but it was hoped that the advantages of composite working would begin to be realized within a few weeks. No one expected the teething troubles to last seven months, or that eleven months would elapse before the planned level of output was regularly maintained.

THE TEAM AND ITS ARRANGEMENTS

One of the conditions of this kind of agreement was that the men should make themselves up into sets of the required number. In the present case this condition was waived by both management and lodge. The colliery was in process of reorganization and a number of putters were becoming redundant, as were men engaged on hand hewing in single places. Management and lodge agreed to draft these men into the new team, together with those who had been engaged on the development of the drift. With the addition of one man who had been on light work, they made up the team of 51 (*Table 21*).

The six spare men were all qualified to pull; one came from pulling elsewhere in the pit, and the other five from composite shortwalls, where two had been deputies.

A meeting of the team was held by the lodge a few days before the start to acquaint the men with the agreement, to allocate them provisionally to tasks and shifts, and to appoint team captains for different task groups. The chairman stressed in his remarks the merits of task continuity and a common note; the drift must succeed, he said, to preserve these advantages. Eight men volunteered for pulling, and ten for stonework; the remainder being allocated to hewing. Three team captains, one for each task group, were elected. Several men expressed anxiety lest they were condemning themselves to permanent nightshift and received assurance that after a week or two it would be possible to start rotating shifts. Arrangements for tasks, shifts, and team captains were, it was stated, all provisional.

TABLE 21 TEAM COMPOSITION

Previous Work	*No.*	*No. qualified as pullers*[1]
Putting	21	4
Hand hewing	12	7
Drift team (a)	2	9
Drift team (b)	5	
Drift team (c)	4	
Drift team (d)	5	
Stonework	1	0
Deputy	1	0

[1] All the men were qualified in the operations of coal-getting and filling, and in stonework.

Representatives of the lodge, together with the three team captains, then met with management to agree final details. Although there had been a vague expectation that three shifts would be worked, it was decided to have only two, concentrating coaling on the dayshift (9.00 a.m.–4.30 p.m.) and doing pulling and stonework on the night (4.30 p.m.–12.00). This gave economies in datal labour and fitted in with transporting the coal from the screens by road. The 33 men on the hewing shift were to be deployed 16 to each face, with one man in the mothergate, while on the nightshift there were to be four pullers to each face, two stonemen in each tailgate and three to each caunch in the mothergate. There were three deputies, all of whom came from cutting faces—including a composite longwall. In charge was an undermanager recently appointed to the pit.

THE FIRST WEEK[1]

Before going in on the first morning, the men on the hewing shift allocated themselves to places by cavilling. As only 29 were present, the three spare men belonging to their shift were called on and also drew places. When the timbering rules had been explained, work began.

Of the 29, 18 had been putters, 5 hand-hewers, and 6 development men. The three spare men were all composite workers. All the putters were new to longwall working, apart from their experience on the training face; they were also new to pneumatic picks. The hand-hewers had no recent experience with pneumatic picks, and little, if any, of longwalls. Only the development men had used pneumatic picks in their previous work. Most of them had also been on longwalls.

The panel had been standing ready for eight weeks but appeared to be in good condition though pools of water had accumulated, particularly on the East face, and there was a 2″–3″ layer of soft clayey stone on the floor. Output had been planned on the basis of a 4′ 6″ web from each hewing shift. The lack of hewing experience, however, soon began to show and it became obvious that 4′ 6″ was not going to be cleared. Some of the men suggested aiming for 3′ 0″. This was also the view of the undermanager, which the team captain passed on, but by the time some men had changed their aim the matter was academic. At the end of the shift there were 20 yds of coal left on the East face and 10 yds on the West, and pulling could not begin.

When the nightshift arrived the deputy sent the eight pullers and two of the stonemen to take the coal off the East face and the remaining eight to take it off the West. By the end of the shift the belts had been moved forward and recoupled, but extreme difficulty was encountered in drawing off; only 27 chocks were taken on the East face and 10 on the West (there being 70 on each).

Tuesday was completely devoted to drawing chocks and re-setting timber. By the end of the nightshift both faces were ready

[1]For a summary of the developments described in this and the next two chapters, see *Table 22*, page 214.

for hewing and a new start was made on Wednesday with a 3′ 6″ web, but it was not completed on either face and the situation once more got out of hand. The pulling task became too much for the nightshift under deteriorating conditions and it proved impossible to maintain orthodox timbering on the dayshift; supports had to be inserted where necessary and reset without regard to position. Stone came down around the chocks making them difficult to trip. By nightshift on Thursday they had sunk some inches into the soft floor and most of them had to be dug out. Often there was so little height that places had to be dug before they could be reset.

The dayshift on Friday started hewing on the West face with all but seven of the men, who resumed pulling on the East face, but these tasks were not completed and by the nightshift the East face roof had lowered to 18″ and frequent falls were occurring on the West. In the early morning of Saturday, 20 yds closed completely and it was decided to abandon both faces and win them out afresh.

The manager had visited the face on the Tuesday dayshift, explaining to each man in person the need for careful setting of timber and outlining the immediate requirements for stabilizing the situation. From this point on he retained direct control. On the Wednesday morning (after the undermanager had left the deputies at the meeting point) he telephoned to aim at an advance of 3′ 6″. At the beginning of the nightshift, having been told how things stood, he ordered the concentration of effort on one face. Managerial attention, however, was no substitute for experience in a team confronted by bad conditions.

There was confusion over team captains; three had been elected when some of the men, including the pulling captain, had thought there were to be three shifts. The pulling captain, in fact, came in on the hewing shift. On Wednesday the hewers elected an additional captain for the West face as the first was on the East. The new man was a reluctant incumbent. He had little authority with the younger members and felt that someone with more experience would have been a better choice. Neither he nor his opposite number had recently been on longwall faces. As for the stonework captain, because of the derangement of work on the nightshift and his own unfamiliarity with pulling, his role in placing the men and co-ordinating their activities was taken over by the deputy.

The conditions encountered proved too much for a team which had no experience of organizing itself and the majority of whose members were unused to each other, to longwall working and, indeed, to facework—especially pneumatic picks. The enterprise represented a miscalculation by management and lodge alike, yet, as the next chapter will show, it was not perceived as such. This raises far-reaching questions concerning what was being left out of account in the expectations of competent and experienced mining people.

CHAPTER XX

A Second Attempt gets into Difficulties[1]

RE-ASSESSMENT AND REINFORCEMENT

On the Sunday following the closing of the panel two representatives of the team, both members of the lodge committee, visited the manager to discover his intentions about payment. He had already decided to pay a day wage rather than hold to the agreement.

It required a fortnight's work to re-win the faces, which was done 16′ 0″ in advance of the old line. During this time two main ideas were discussed among the men for improving the situation: closer supervision of timbering, with a man appointed to each face with the sole duty of supervising the inexperienced, and some rotation of shifts as soon as conditions permitted so that the less experienced could see how the work of each shift affected that of the other.

The manager held a meeting with twelve of the men at which several decisions were made. Supervisors were granted for each face on both shifts; an extra deputy was allocated to pulling; and a token third shift was arranged, eight of the hewers going in on the foreshift to 'break in'. Back caunch men were to come in two hours later, since their work was impeded when the cycle lagged. Diagrams of the method of timbering were issued and work began on the Monday in a mood of cautious optimism.

The pattern of advance in the first week of the new phase, and in the six subsequent weeks until the next crisis, was one of concentration on alternate faces each day. This concentration was not exclusive, since some work was done on the less favoured face. Pulling was concentrated on the face which lagged least and the next hewing shift then concentrated on the other face. During this period conditions remained bad, particularly on the

[1] Source paper: 17.

West face. Towards the tailgate, chocks had to be drawn with sylvesters. Operations were hampered by ramble, which fell between the straps and which, since there was no space in the goaf, had to be stored on the face and run off at the end of each shift. Foreshift activities were concerned with clearing up and reducing lag as much as possible on one face.

After his meeting with some of the men on the Saturday preceding the resumption of work, the manager felt that everything had been done on his side but that the team's internal organization could be improved. He did not know who the team captains were. He thought the original men had given up because they did not want the responsibility. He had advised captains for each face on each shift, but doubted if they would find them. Nevertheless, he expected things to go better, ascribing previous difficulties to bad timbering rather than bad conditions, and to the fact that the men were not working as a team.

In the team itself the two hewing captains both transferred to the foreshift. Two possible successors were mentioned but only one took any active part, that of representing the team in later discussions with management. The deputies were vague about who the team captains were supposed to be and they (and some of the men) began to joke that members took turns at the position every day.

According to the deputies, the differences in effectiveness among the men on the hewing shift were too great to continue much longer. The more effective would finish before the end of the shift; they would then be asked to move, or would move voluntarily, to another place; there they would see how little had been done by the less effective and their reluctance to do more themselves increased. The men expressed similar feelings. In only a few cases, however, were members of the team regarded as unwilling to do their share, 'better' reasons usually being ascribed for poor performance—that a man was inexperienced or unskilled, or that, being young and unmarried, his wants were few and his incentive correspondingly low. Feelings nevertheless existed that some men were not pulling their weight.

This applied not only to the quantity of coal hewn but to the quality of timbering. The style favoured by the supervisors differed from that favoured by the deputy and this led to confusion among the less experienced. Timbering which gave firm support at first became unreliable once the ramble above it began to break up. These problems gave rise to tension between, as well as within, shifts.

The supervisors allocated to the hewing shift were selected by management from a team of bargain men, who had been working alongside the team though not paid out of its earnings. Officially their task was solely instruction but they were subjected to pressure to reinforce the hewing. Instruction suffered to such an extent that little progress was made in educating the putters in the use of pneumatic picks.

On the nightshift, among the pullers and stonemen, things went rather better. The pullers, having decided to continue in the sets in which they had begun, cavilled to decide faces. The stonemen resumed work in their previous places, with the exception of two who had moved on to the dayshift and whose places were taken by spare men. These moves were unilateral and there was general uncertainty about the locus of authority for shift changes. In the second week these two men returned to nightshift and one other exchange was made. At this point a demand was voiced that cavills should be drawn for places, though no one felt certain whether pullers and stonemen should be cavilled together, or whether the cavilling should be done individually or by pairs. Those who voiced the demand wanted a means of rotating places from day to day as some were decidedly less pleasant to work in than others. Nothing came of this for several reasons: not all the stonemen were qualified to pull; the appropriate groupings varied in size; one group of stonemen came in at 6 p.m. and their opinions had not been heard; and the duration of spells on nightshift was uncertain since no settlement had been made about rotating shifts. The deputy said that, though he was willing to have the cavilling carried out, he would rearrange the men as he thought necessary in the interests of the work. The matter was allowed to drop.

On another occasion when management wanted to pull only on one face, the deputy decided that six men should do this while the other two helped the stonemen. He called for volunteers. An objection was raised that one of the volunteers came from the face that was to be pulled and that the correct procedure was to have volunteers only from those pullers who did not have their 'own' work available. This point of procedure was agreed by all the men and carried in face of the deputy's opinion that it was irrelevant on a composite face. Another point of procedure was raised by a stoneman. It had been necessary to bring some men in on a Sunday, to work in this man's place. He raised the point, explicitly as a procedural one, that, although he would not have come, he should have been the first to be asked since it was 'his' cavil.

The variation in the ability of the men on the hewing shift had its effects on the nightshift. Apart from taking off such coal as remained, the pullers had to reset a good deal of timber. This added to the delays caused by having to clear the faces of stone and to the labour of chock-drawing, particularly on the West face where it was still necessary to use sylvesters. In these circumstances the men on the nightshift wanted a change. On the Thursday of the first week of the new phase one of the pullers went to the meeting-station and told each of the hewers as they came out that the pullers intended to have the cavils put in for shifts at the lodge meeting on Sunday. Next day there was some dispute between hewers and nightshift because some of the latter had been working overtime and payment for this came off the general note. The nightshift (and the deputies) pointed out that overtime had been necessary to keep the face going at all and that the hewers would change their views when they experienced conditions on the nightshift. Anonymous voices from the background replied, in the heat of the moment, that they couldn't go on the nightshift because they weren't qualified to pull.

THE ISSUE OF RECONSTITUTING THE TEAM

It was with some knowledge of this background as seen by the deputies and undermanager that the manager met the team captains on the Saturday. He asked them to dissolve the existing team so that the men could choose a new one from among all those working in the drift. Would they meet him again on Monday after discussing his proposal at the lodge meeting? The agenda for the lodge meeting was full and no time was found. A message was sent to the manager, who had also heard of the intention to cavil for shifts. In reply he insisted on experienced pullers staying for the time being with their task. If cavils were drawn for shifts, and inexperienced men allocated to pulling, he would stop the face and disband the team. The lodge decided that two men should attend a meeting on Wednesday between the manager and the team captains.

At the Wednesday meeting no agreement was reached. The manager repeated his view that dissension was hampering progress, adding that he knew that some men wanted to come off and that there were others whom the team would like to drop. Would they settle this? For his part, he was willing to provide two extra hewers on dayshift and two extra pullers and one stoneman on nightshift, to be paid for by him and not out of the common note. Union officers visited the face for a 'pep talk' with the team, but decisions were left to the lodge meeting the following Sunday.

At this meeting the chairman outlined the events of the previous two weeks, and gave the manager's views. The manager, he said, seemed reasonable; the men, for their part, had a responsibility too—'we have to put our own house in order.' The formal purpose of the meeting was to consider the manager's suggestion that the team reform itself, but this topic, with its implication of discarding members, was put aside. The chairman first presented the situation in its wider setting as seen by the lodge committee, there being three considerations: the nature of the agreement; the importance of the drift to the life of the pit; and the present position as a test of the new type of agreement.

The advantages of the composite type of agreement were contrasted with the disadvantages of the old 'specialist system' with segregated work groups, divided paynotes, and a status hierarchy with a corresponding hierarchy in rates and earnings. Under the composite system, though individuals would be better at some things than others, their special skills would complement each other, so that a man could 'find his niche' and make his contribution without having to accept a lower rate of pay because of the task he performed. Moreover, the new type of agreement eliminated the need for earnings being made up. There was an end to one group or individual being favoured or getting special treatment. There were no 'glamour boys' and no 'Cinderellas'.

The colliery was a 'village' pit in a coalfield that was beginning to die out. They had all been through the crisis in its economic life some two years ago. It was as part of the plan to meet this that the composite method of working the longwall had been introduced. Although this had succeeded in another seam, with cutting faces, it was still on trial and the new hewing venture was crucial.

The chairman then moved to the question of team changes. After reminding them that the drift had been working for a comparatively short time, he repeated his previous point about men being better on some tasks than on others and stressed the need to avoid hasty judgements about people's worth. Even where men were almost certainly not pulling their weight, it was hasty, if not harsh, to throw them off the team 'without further ado'. They should first be spoken to reasonably and an attempt made to find out what was wrong, so that final action would be taken only in the case of 'incorrigible drones'. To throw a man off the team was something not to be done 'impulsively', because the man would carry a certain stigma for the rest of his life in the pit. Where a man was trying hard but was not up to standard, he had to be given some consideration because he too had his living to earn. In the case of men who wished to come off the team of their own accord, there was nothing the union could do to stop them, but they had to remember that those who left the team would have to take what jobs could be found for them elsewhere in the pit and such jobs might not be particularly well paid.

He now asked for the names of any who wished to come off. The first name given was that of the stonework team captain. Two other stonemen gave their names, and were followed by a puller and a hewer. After a pause one of the younger men, an ex-putter who had originally been elected team captain for the hewing shift, but who had relinquished the role, said something more had to be done. There were people on the team who weren't pulling their weight and ought to be off, and they weren't those who had just volunteered. The fact that this feeling had been expressed was welcomed by the lodge committee, but once again the chairman emphasized that they ought not to be in a hurry about such things.

Two further points were raised. The first was technical. One of the pullers suggested that the hewers might be more careful about setting straps. The second point concerned shifts. Some of the nightshift men protested that they had volunteered for nightshift on the understanding that this would be temporary. Nothing came of this complaint. The manager had vetoed any changes for the time being and the lodge committee's view was that, to make a success of the drift, changes should be postponed until things were going properly. This was accepted without enthusiasm.

At times during and after the meeting the 'dissension' in the team was mentioned. Several of the pullers and stonemen remarked that they knew of no dissension among the nightshift. Any dissension was confined to the hewing shift. This was also the management view. When composite cutting faces in the Manley were referred to as a model of good relations some of the younger dayshift men retorted that they were getting tired of having these examples thrown at them.

On the question of dropping men, some of those present who came from other parts of the pit expressed surprise that inequality of contribution should be regarded as something to worry about, citing their own teams as examples, and saying that the range of output among the men varied widely but nobody thought anything of it. It was pointed out that in those teams the men had chosen each other and had therefore accepted these differences in advance, whereas on this team the men had been placed together.

Five men left the team and were transferred. The youngest was 44, the next 45, and the other three, stonemen from the mothergate, were all 58. The management expressed the view that they had left because the younger men had been reproaching them with failing to keep up. This was unlikely as four were on nightshift, but is an instance of management's perception of the nature and causes of the dissension. At the same time two of the younger men from the dayshift (aged 23 and 26) left the pit and the industry and another went off sick because of an old injury, returning only to light work. The numbers of the team were thus reduced by eight, six of whom were replaced by the six spare men in the drift. These latter were not replaced. Though volunteers were called for, none was forthcoming.

The changes which the manager had hoped for had not come about. He was disappointed in the actual changes, believing as he did that some of the older men had been coerced into leaving. He thought these departures would weaken the team by reducing the body of experience available.

CHAPTER XXI

Crisis and Resolution[1]

THE MANAGER ENFORCES THE AGREEMENT

For the three weeks following the lodge meeting things remained quiet, mainly because of the Christmas and New Year holidays, though two more men left the pit and the industry and informal shift exchanges began. A man wanting to change his shift had to find someone on the other shift with whom to change. This usually meant canvassing. Nightshift men would stand at the meeting-place and ask for swaps with the dayshift as they came out. Some of the ex-putters were accused of monopolizing the dayshift and of capitalizing on their lack of experience to stay on it.

Though within the hewing shift the feeling remained that certain men were not doing their share, this feeling was never directed at specific individuals. Some men, it was said, must be taking 'short ground'; each man should mark off his place or the deputy should put chalk marks on the coal or, better still, on the air-pipe. The deputy pointed out that the face had not been broken into at regular intervals and that this may have produced a false impression. That stint marks should have been suggested in a pit where they were regarded with contempt is an indication of the strength of feeling aroused. Once the immediate anger had subsided the suggestion was discounted.

The first full week after the holiday period (week 9) began quite well and on the Monday the hewing shift, with two men short, cleared off the 4′ 6″ web with the exception of 20 yds—though they were not able to keep this up. Conditions had improved somewhat on the East face, where chock-drawing had become relatively straightforward, though there was still a good

[1] Source paper: 17.

deal of water. On the West face, there had been little improvement and towards the tailgate conditions were as bad as ever. Among the team there was a wide variation of opinion about the rate of progress to be expected, with less optimism among the pullers than the hewers.

The manager again had a talk with the team captains: things were not going well; he had done all he could; from now (week 9) on he intended to pay strictly in accordance with the agreement. Disappointed that none of his efforts had had much effect he believed there was still dissension in the team which the men would not admit and that this was the main cause of lack of progress. Following this announcement of his intention to pay in accordance with the agreement, he visited the face, criticizing the timbering, which was bad in places, and putting his foot down on finding experienced pullers on the hewing shift. His insistence that three of these men return to nightshift caused particular dismay.

Towards the end of this week, and more so in the next, conversation centred on the level of pay to be expected under a strictly interpreted agreement, estimates of five to ten shillings lower per shift being given. There were rumours about people leaving, though no one could ever say who, and only one man gave notice. The others, as the Thursday of week 10 drew nearer, contented themselves with speculating about other jobs. The present level of earnings was little enough, in their view, for the work involved under such bad conditions and anything less would not be worth it. There was also some doubt about whether the agreement was enforceable, given the nature of the conditions. Only a few of the men were familiar with the details, and they began thinking they should all have copies. They changed their attitude to management: the manager was now regarded as obstinate in his general attitude and rigid in his insistence on adhering to the original plan; as to the undermanager, 'things have got beyond him'.

Meanwhile, conditions were once again deteriorating, particularly on the West face. The combination of friable roof and soft floor made methodical timbering difficult, and chock-drawing again became a protracted operation involving digging and the use of sylvesters. The pattern of advance was still that of concentration on alternate faces with the aim of getting as much coal as possible off one during the dayshift without losing control of the other. On the Wednesday nightshift (week 10) the manager inspected the West face and decided that conditions were so bad that he would withdraw the men and shorten the face. By allowing the western half to close, he could concentrate on rectifying the support of the eastern half—the 40 yds nearest the maingate.

The deputies expected things to come to a head on the Thursday when the men found how much their pay had dropped. When they received their notes they found average earnings down by 30/-. They went in a body to the office and asked the manager to see four representatives—one from each face and shift. The manager agreed to see three. He refused to discuss paying more and insisted on sticking to the agreement. The shortening of the face meant a reduction in manpower; and he asked them to make up a set of 42 from all the men available in the drift. He wanted only experienced pullers on the nightshift for at least the next fortnight, after which they could think about rotating inexperienced men to learn the job. He asked the three men to see him again the next afternoon, hoping that the shortened face would provide the opportunity to get the nine worst men off the team.

To the men, picking a team of 42 or choosing nine to be dropped was the same and could not be allowed. There was a procedure in the cavilling rules for handling such circumstances. To comply with the manager's request amounted to picking men to be dropped from piece-work. The only way open to the manager to achieve his aim within the rules would have been to take the whole set off the face and then to ask for volunteers, of which none would have been forthcoming. As to his view that less work was being done than could be expected, they believed he had misinterpreted the circumstances and had failed to allow for the conditions. The pullers pointed out that two men might move only six chocks in a shift but that to put it in those terms overlooked the work entailed in digging out and resetting. Despite his visits to the panel and his action in shortening the West face, they felt the manager was unaware of how bad conditions had become, especially for the nightshift.

Before the three-man delegation met him again on the Friday, one man was asked to stand down, as he was thought too prone to agree with the manager. His place was taken by a putter, since 'there were 21 putters on the team and nobody to represent them'. The manager was now told that none of his requests was acceptable and that the whole situation would have to be discussed at the lodge.

THE LODGE ATTITUDE

At the lodge meeting on the Sunday (beginning week 11) the secretary said that the manager had informed him that the West face had been shortened and that it had therefore become necessary to reduce the size of the team. The chairman said that the men in this pit had always been opposed to anything which savoured of 'survival of the fittest' or anything which could lead to one man being preferred over another 'because someone liked the colour of his eyes or the way his hair was parted'. The cavilling rules had been built up to deal with this and to cover situations of the kind which had arisen. They had to be followed, despite the manager's previously and frequently expressed opinion that the cavilling rules should be 'thrown out of the window'. The manager was getting tough and putting the blame on the men. Although 10 per cent of the responsibility lay with the men, the manager had been told about conditions but had refused to take notice. If he continued to ignore the facts, the lodge would seek a meeting with higher management. If the manager refused to arrange this, they would declare a dispute and such a meeting would follow automatically. A fortnight ago the manager had been delighted with the Monday achievement of removing all but 20 yds but had refused to listen when the difficulties of pulling were described. There was nothing in the agreement which compelled the team to attempt a web of 4′ 6″; if they felt 3′ 0″ was more within their scope and would give more time for pulling then that was what the team captains should decide. Also, work such as digging chocks out and digging to get them in again came outside the scope of the agreement. Some of the men raised similar points, such as having to build wooden butts to make pack walls because of a shortage of hard stone. These points were noted as material with which to approach the manager.

It was then moved and seconded that cavils should be drawn to decide who should come off the team. The chairman thought that the manager had rushed them and proposed the actual procedure be left over until the next meeting. This was agreed. One of the men remarked that the manager seemed to know a good deal about what went on in the team and that, although he heard things officially from the deputies, some of the men also must have let things slip. The chairman observed that no reasonable man would mistake things said in the heat of the moment for considered opinions, but the manager was not a reasonable man. It was agreed that no one should see the manager unless a member of the lodge committee was present.

In the following week (week 11) the lodge committee met the manager without reaching agreement. Nevertheless, the manager felt that at least each side had stated its case and had disagreed without 'getting at each other's throats'. During this week the Area Labour Relations Officer made an unofficial visit to the pit and looked around the drift. There was a further meeting in week 12 between the lodge committee and the manager when the question of pay was again raised and again refused, but the manager and members of the committee agreed to visit the drift together to see both hewing and pulling. A lodge meeting was held to report progress, the committee announcing that if they were 'unable to get any sense out of the manager' they would be prepared to withdraw labour from the drift, 'even if this meant jeopardizing the output of the colliery'.

During the three weeks of negotiation output was low. In week 12 some of the pullers decided that they had had enough of continuous nightshift and unsuccessful canvassing and put the cavils in for a change of shifts. That the manager had insisted on leaving experienced pullers on nightshift was discounted; he was not paying enough; they were being imposed upon and had therefore no obligation to stay on nightshift. More resentment was expressed than ever before that some men on dayshift had never been off it and never would be unless a formal procedure prevented evasions. Some dayshift men queried the validity of the cavils but were challenged to bring their queries to the lodge meeting. The deputy agreed to enforce the cavilling allocations by sending home anyone who turned up on the wrong shift.

In the first week of the crisis the feeling about the level of earnings had been one of resentment, mixed with surprise that the manager had chosen to enforce the agreement in the week during which output had been the highest. It was, they said, ironic that the highest output should have produced the lowest pay. The second week's pay was lower still. When this became known the mood changed from resentment to dismay. In the third week, despite earnings which were lower still, the atmosphere became more cheerful. The change, remarked on by both men and deputies, was attributed to the fact that negotiations, however difficult, were still proceeding. This was taken as a sign that a settlement was likely to be reached.

THE NEW AGREEMENT

In week 13 the proposed visits by the manager and members of the lodge committee were made to both shifts and another meeting was held at which the Area Labour Relations Officer was present. Agreement was reached on the following points:

(a) The rate of payment for output above the basic two cubic yards per manshift was increased to an acceptable figure and the manager agreed to make up the wages for weeks 12 and 13 to the level that had obtained before he enforced the strict agreement.

(b) The web attempted was reduced from 4′ 6″ to 4′ 0″ to enable the men to accept the task as within their compass.

(c) The manager made a list of the men he wanted on each shift and it was agreed to follow this for two or three weeks.

(d) Although the manager had withdrawn the extra men from the face when he enforced the agreement, he now agreed to put two extra men, paid by himself and not off the team's earnings, into the tailgates where the work was falling behind.

(e) Four men were cavilled off the team (since the team had already lost five because of sickness or transfer to other industries). Contrary to the manager's original intention these four stayed in the drift as 'spare men'.

This agreement was endorsed by the lodge meeting on the Sunday and work on the Monday of week 14 went noticeably better.

THE EFFECTS OF THE SETTLEMENT

On the first shift under the new arrangements all the coal was taken off and the air-pipes and belts moved over into their new positions. Though this was not repeated for both faces for several weeks, the rate of output improved until in week 17 it reached the level planned for the shortened face. In week 22 a start was made on opening out the West face and 10 weeks later this was completed, output just exceeding the 1,000 ton mark. The scheduled target, however, was not consistently maintained until after week 44.

For the first week under the new agreement (week 14) pay exceeded, though only by one shilling per shift, that for the two previous weeks when it had been made up. Some of the men thought such a difference too small for so marked an increase in output, but the team generally did not accept this as a valid point. Before the actual earnings became known there was a good deal of speculation about what they would be, expressed in the form, 'I wonder what he will give us', as if somehow the figure depended upon the goodwill of the manager.

In week 16 some of the men began to exchange shifts on an individual basis, but there was no system of shift rotation over the team as a whole. On the Monday of week 17, 25 men appeared on dayshift instead of the scheduled 19 and there were 6 men short on nightshift. The undermanager and the deputies regarded this as proof that the team could not manage its own shift changes and took over the function themselves, designating the men they wished to see on nightshift and announcing that anyone coming on the wrong shift would be sent home. In the following week a blackboard was put up at the meeting-place, on which men entered their names for the next week's shifts. The deputies, with the undermanager, rearranged such men as they thought necessary, on the principle that regular nightshift men should be given a spell on dayshift and that less able dayshift men should be brought into the nightshift to do stowing in the tailgates. In addition, dayshift men with no pulling experience were brought into nightshift every other week to complete the statutory period of training. At this time there were 19 men on dayshift, 18 on nightshift, and 4 on the foreshift. Since the latter were volunteers, it was a simple enough matter to allow everyone to alternate night and day.

After the extension of the West face, new men were fed into the team. This complicated attempts at equitable shift rotation and management retained control on an *ad hoc* basis. Once the face had been extended to its normal length and the team built up to full strength, the rotation of shifts became more systematized. Eighteen men, in three groups of six, rotated weekly over all three shifts, while the remainder alternated two weeks' dayshift with one week's nightshift. Within this broad pattern there was room for private arrangements among individuals. By this time the drift had settled down to the regular production of 1,000 tons per week.

Things remained at this level for six more months. There was then a drop in coal height of 7 per cent and a consequent reduction of output. There was also pressure from the men for a change in price. Following a precedent from elsewhere in the pit where better prices had been obtained for single units, the manager offered to treat the panel as two separate faces. He still felt that there was some dissension among the men and that such a change might help. The offer was accepted. The team split into two by mutual selection and work on the new basis began, 73 weeks after the opening of the drift.

CHAPTER XXII

Underlying Forces and Group Defences—An Analytical Commentary[1]

THE ASSUMPTION OF ORDINARINESS

This account has been a chronicle of how things went wrong. Management, lodge, and members of the team all assumed that the drift would be an ordinary unit, to be run in the ordinary way and unlikely to experience more than ordinary difficulties. The untenability of this assumption, yet the persistence of behaviour based on it, suggests that, in a way they were unable to recognize, those concerned were using the idea of ordinariness as a means of psychological defence against elements in the situation they were unwilling to confront. The principal effect was that the panel was treated throughout as a production unit under difficulties, rather than perceived for what it was—a training and development project working under the stress of a demand for full production. Though the need for support was conceded, the time allowed (eight weeks) was no more than a token period and the underlying assumption manifested itself in a number of ways, especially in the emphasis on coal production as the primary task. Though, in words, the need for training and acquiring experience was acknowledged, preoccupation with output carried the real message of action.

The degree of unreality in assuming ordinariness may be measured by the fact that an inexperienced management and an inexperienced work team were brought together in a novel task under conditions likely to be difficult—especially in the beginning. The undermanager was new to the pit—it was, moreover, his first appointment as an undermanager; the deputies were new to each other and to the men, and the majority of the men were new

[1] Source paper: 17.

to any form of longwall working. Though no one knew the problems likely to arise in developing a hewing panel on composite principles, there was ample precedent for anticipating trouble if the scatter of hewing abilities was too wide. Again, though no one could anticipate in detail the conditions likely to be encountered in the drift, plenty of experience was available in the neighbourhood to suggest that the roof was likely to be awkward until the faces were further in from the outcrop—which meant that pulling experience would be at a premium.

All this was known by management and lodge when they entered negotiations. It was known also by the men. Yet it was all disregarded. The pit was under severe pressure to become economic—as soon as possible—and there was a great deal of anxiety about this; hence the preoccupation with production. The Manley composite panels had been successful—beyond expectation and without trouble. The real effort, however, in the original negotiations had been on the terms of the agreement. Problems of work organization had been left to take care of themselves. This they had done, in a very remarkable way, but no analysis had been made of the reasons; it was simply taken for granted that such problems would take care of themselves again. Meanwhile, reorganization elsewhere in the colliery had made redundant a considerable number of faceworkers. The lodge sought conditions for their redeployment without the relegation of any from facework status. Moreover, the one piece of 'hard news' which had circulated about the Manley was that earnings were high—higher than earnings had ever been in the pit. It was an attractive prospect, therefore, to workers finishing odd jobs in older workings, and wondering what might happen to them, to join a new drift and get in on the advantages of the new type of agreement.

This led to the one feature which was not ordinary about the drift: the team was drafted, not self-selected, with the result that there was no commitment to accepting the differences in skill and experience which were later discovered. Even the safeguard inherent in the traditional procedure for forming teams was dispensed with. One may infer that the pretence that special measures were unnecessary and that unusual risks were justified covered a pervasive fear that the drift would not succeed, with the serious repercussions this would have for the future of the colliery. This is the underneath anxiety, intense in the change situation, which brought about the collusive denial of reality which in turn led to the initial assumption of ordinariness.

THE REACTIONS TO FAILURE

The reactions to what transpired after work began may best be followed from *Table 22* (pp. 214–15), which relates the events described in detail in the previous three chapters to the amount of activity induced in wider managerial and negotiating systems and to the levels of reinforcement, leaving, and productivity which characterized different phases in the socio-technical history of the drift.

One might have supposed that the closure of the face at the end of the first week would have given a big enough shock to cause a radical re-appraisal of the whole undertaking. Confrontation of what had happened, however, at a higher level of reality would have meant giving up the assumption of ordinariness. Instead, during weeks 4, 5, and 6 its role as the 'chosen' defence was bolstered up by a series of measures that attempted to make it work in defiance of the facts. This is the latent meaning of the generous reinforcements offered by the manager (equal to one-fifth of the team's strength) and of the extent of his support by the lodge. Nevertheless, a split in attitudes and relations was already detectable: at the same time as giving reinforcement, the manager asked the team to reorganize itself in a way which brought it into direct collision with the cavilling rules; and the men, perceiving this as an attack, started to go into opposition. This is the negative side of the collusive process, just as much lacking in task orientation as the positive side. It led into a type of impasse, familiar in industrial relations, in which no learning takes place from what is being experienced.

As the illusion wore thin and the impasse became more apparent the men began to despair (by week 9 a quarter of the team had left), while the manager, desperate over costs and production and interpreting the men's lack of response as an act of hostility, enforced the agreement both as regards pay and manpower. This sudden regression to coercive control—punishment-centred

TABLE 22 DEVELOPMENTS IN THE DRIFT

P	*W*	*Events*	*M*	*R*	*L*	*T*
1	1	Difficulties too great for unorganized and inexperienced team. Face closes under extremely bad conditions.	–			<20
2	2	Rewinning.	–			0
	3	Rewinning. Manager meets 12 of team: reinforcements granted—4 technical supervisors, 1 extra deputy (pulling); production target reduced—alternate face concentration; token third shift of 8 hewers to break in; make up of wages on a day basis.	1			0
3	4	Poor production progress—conditions still bad. No internal team organization; no permanent team captains; differences in hewing ability cause trouble; face supervisors pressed into working rather than instructing. Manager meets team captains: wants team dissolved and new team picked, eliminating poorest workers.	1	5		<50
	5	Manager meets 2 representatives of lodge committee and team captains; no agreement over dropping poorer workers. Manager offers further reinforcements: 2 hewers, 2 pullers, 1 stoneman.	2	10		<50
	6	Lodge meeting reviews situation, chairman supports manager, holds men back on premature demands for shift changes, warns against hasty judgement over dropping men. 5 volunteers come off (older men).	1	10	5	<50
	7	3 younger men leave. 6 spare men make up team, are not themselves replaced. Volunteers called for, none come forward.	–	10	8	<50
	8	Christmas holidays. 2 more men leave.	–	10	10	<50
4	9	Manager loses patience with poor production and inability of the men to sort themselves out. Enforces Agreement. Withdraws extra labour. 2 more leavers.	1	0	12	<40
	10	Manager shortens west face by half in view of persistent bad conditions. Only 42 men now required. He asks for new team, dropping 9 poorest. Delegation of men refuse to make such judgements; want cavils. Pay under strict agreement drops by 30s 0d per week. Shock and	2	0		<30

P	*W*	*Events*	*M*	*R*	*L*	*T*
		resentment. Manager refuses to pay outside the agreement.				
	11	Lodge insists on cavils over the 9 surplus men. Previous support of manager replaced by opposition. Proposal to seek meeting with higher management. If refused by manager, lodge would declare dispute—such a meeting then automatic. Manager and lodge committee meet—no agreement. Area Labour Relations Officer visits pit.	3	0		<30
	12	Manager/lodge committee meeting: manager refuses to do anything about pay but agrees to visit day and night shift with representatives of lodge. At subsequent lodge meeting, committee announces intention of withdrawing labour from the drift if manager won't come to terms.	2	0		<30
	13	A new agreement is reached between management and lodge with the help of the Area Relations Officer. Wages improved, target reduced from 4′ 6″ to 4′ 0″, redundant men cavilled off (only 4 left out of 9). Special men to remain in key roles on each shift until difficulties overcome. 2 men granted as reinforcements.	2	0		<30
5	14–17	Face management steps in and creates rotation system. Planned production reached, proportional to shortened face, within three weeks of the settlement.	–	2		<70
	18–21	Consolidation of team.	–	2		<70
6	22–33	West face opened out. 1,000 tons per week (planned target) reached for first time in week 33.	–	2		<90
7	34–43	Fluctuation about planned target.	–	2		<100
8	44–72	Steady state 1,000 tons per week.	1	0		=100
9	73–	The two faces go on separate notes as better prices for single units had been obtained elsewhere in the pit.	–	0		=100

P = Phase; W = Week; M = Meetings (management/lodge, lodge); R = Reinforcements (cumulative); L = Leavers (cumulative); T = per cent production target bureaucracy in Gouldner's terms (1955, p. 207)—produced the corresponding stereotype of militancy in the men's reaction, and manager and lodge found themselves in head-on conflict, with the threat of a dispute.

In the terms introduced by W. R. Bion (1950) for the description of unconscious group processes, *basic assumption fight-flight* (*b a f*) had been mobilized and suffused the behaviour of the group, management and workers fighting each other in common flight from the problems that had to be solved in the real task situation. In week 12 the mood on both sides changed after the visit of the Area Labour Relations Officer. In Bion's terms, basic assumption fight-flight had now been replaced in the emotional life of the group by *basic assumption dependence* (*b a d*), and in this modality a settlement was reached with the help of a 'wise and benevolent' figure representing the higher authority of the Area General Manager—an extremely 'good object' to everyone in the pit. Within three weeks there was a dramatic improvement in productive performance with the target reached in proportion to the shortened face.

The new agreement recognized more of the realities of the situation than had the original and represents a partial undoing of the assumption of ordinariness, some learning through experience having taken place.[1] On the other hand, the working group continued in a management-dependent phase for several months with the deputies stepping in and making all arrangements for face deployment and task-shift rotation. This had a reality component in that new skills and relationships could not be fully consolidated until the face was opened out again and the team built up to full strength. Some outside help on matters of organization was beneficial while task learning proceeded and new members were being absorbed. Nevertheless, it was eleven months before the production target was regularly reached. Even after eighteen months, when the panel split into two teams on separate notes, there was still doubt in the mind of the manager (and the observations of the research team confirmed this) as to whether the drift groups had attained the cohesive independence of the Manley panels. The original collusive denial of reality and the subsequent pattern of interactions between management, lodge,

[1] cf. Bion (1960) on the hatred of learning through experience.

and the working group had impaired, at least for a time, the capacity to develop responsible autonomy.

CORRECTIVE MEASURES

With the situation structured as it was and the process started on the path taken, the subsequent course of events was to a large extent already determined. The available resources were used, the actors in all roles behaving very much according to expectation. As soon as the working group and immediate face management showed themselves unable to contain the situation, representatives of the next largest system (the colliery) stepped in, with the manager taking direct control and the lodge becoming officially involved. When the situation still remained out of hand, the even larger Area system became implicated; and, though a settlement was now reached, matters were only put right in the 13th week after much expenditure of emotion and time, serious losses in production, and the incurrence of substantial additional costs in rewinning faces, making up wages, and remunerating reinforcements. The working group was inhibited from developing responsible autonomy and the whole episode may serve as an illustration that no amount of management from the outside, whether supportive or coercive, can replace effective self-regulation by the primary group.

One may ask what measures not immediately available in the surrounding work culture—since those available were used—might have prevented the situation from developing in such a troublesome manner. To answer this question, one must ask how an assumption of ordinariness could have been prevented in the first place, which is equivalent to asking what might have permitted the negotiating group to provide leadership in the reality rather than the irreality dimension (cf. Lewin, 1935). The view is put forward that members of the negotiating group would have been able to master more of their own anxieties, which represented those of the colliery at large, and so been able to work out a more realistic scheme if there had been a fuller understanding of the reasons for the success of the Manley panels. They would then have known something of the conditions required for effective team work in composite longwall working and would have recognized the serious implications of their absence in the drift. At the same time such a fuller understanding would have demonstrated that it was not impossible to make a plan which would have brought the required conditions into existence. An effective plan, however, would have entailed complete abandonment of the assumption of ordinariness—with its implications that full production was realizable after a brief period of settling in. The drift could then have been set up as a special training and development unit with the prior task of becoming a balanced and cohesive work force as a condition for attempting target production.

Such a step, however, would not have been easy within the norms of the prevailing work culture. These permitted the crisis to be resolved without a dispute and, in the end, a level of production (o.m.s. approx. 5 tons) to be reached which was better than that likely to have been achieved under conventional arrangements. What the prevailing norms did not provide was any precedent, or 'tool kit', for analysing factors in the socio-psychological system in a way which would have broken down the assumption of ordinariness in the starting situation, and avoided the consequent tensions and loss of production.

SECTION SEVEN

Change Processes

CHAPTERS

CHAPTER XXIII

Organizational and Technological Change[1]

THE IDEA OF THE OPERATIONAL EXPERIMENT

During the course of the research, which was at first directed towards the understanding of on-going systems, opportunities were sought to study units in the process of undergoing technological or organizational change. The intention was to identify the nature of the change and to follow its course. A change situation was defined as one new to the pit concerned. The situations studied, therefore, ranged from the applications of methods commonly found in other pits, such as double-unit layout, to major technological experiments or organizational developments, such as the installation of a Haarman scraper-peeler or the emergence of three-shift composite longwalls. Whether or not the innovations were more than local in their novelty, at the pit concerned they had the quality of *operational experiments*. The outcome of each was surrounded with some degree of uncertainty, though this was naturally much greater when something radically new was being attempted. How far such activities were perceived as 'experimental', and the change process approached accordingly, is a matter to which we shall return at the end of this chapter.

THE NATURE AND SCOPE OF THE CHANGE STUDIES

Nine change situations were studied, including those reported in Sections Five and Six. In six the process was followed from its inception; in the others observations began in the early months. In all cases the issues were very much alive in the minds of those involved on both the union and management sides. This section analyses change at the level of partial mechanization: new methods

[1] Source papers: 6, 19, 20, 23.

of working within the conventional system; the transition from conventional to composite longwall. The introduction of higher mechanization is deferred till Section Eight.

As regards change within the conventional system, an account is given in Chapter XXIV of the effects on face and seam organization of a decision to concentrate production by replacing single with double units. The study embraces a change period of 15 months—from the steady state of low production which existed when the whole seam was worked with single-unit hewing longwalls until a new equilibrium was established at a higher productive level with one cutting and one hewing panel. Of special interest is the constructive use made of the single place tradition in the quasi-composite organization of a large, double shift, hewing group; the impasse created by the cavilling rules—removed only as a side effect of technological change; and the merits and limitations of the step by step method of change management.

Chapters XXV and XXVI deal with change to composite organization in a seam worked by a variety of methods—principally conventional longwall. The focus here is on the seam system as a whole, rather than on face group organization, though a 'scale of compositeness' is developed based on the four aspects of composite working described in Chapter IX which allows the degree of compositeness of each face to be related to its productive performance. Just as the previous chapter raises the question of how far a more systematic socio-technical approach would have brought about the same degree of change in less time, this chapter suggests that such an approach would have resulted in a greater degree of change in the same time. In both cases the inference is that increases in production could have been achieved considerably larger than those obtained.

Section Eight presents studies of longwalls on which elements of higher mechanization were introduced. The objective is to ascertain the extent to which existing forms of work organization, values, and customs were carried over to new systems and to assess their degree of appropriateness under the changed conditions and their effect on productivity. The persistence of existing features is also considered from the point of view of resistance to change. A Haarman scraper-peeler unit illustrates how existing production norms for partially mechanized faces continued to determine expectations on a highly mechanized face, though a break had been made with traditional practices in team selection. A new set of work arrangements is then described capable of yielding a higher level of machine utilization. The function of such models is to increase belief in the attainability of new norms.

A longwall on which flight-loading and scraper-packing were introduced shows how orthodox methods of teaching a new technology may inhibit the emergence of more appropriate forms of organization. The vicissitudes experienced in changing from hand-filling to flight-loading on a composite longwall—a technological advance in no way presenting a major learning problem—point to the importance for team building of taking into account the origin and social structure of existing work groups. A social technique for the rapid dissemination of a new technology, which had considerable success over an extensive trial period, is described in an account of an Area training school which the research team helped to develop. This approach sets new equipment in the context of a system of mining considered as a socio-technical whole and is contrasted with the type of Area development team, employed on the two flight-loading faces, which concentrates exclusively on the training of one group of operators in an isolated machine process.

The appositeness of composite organization for higher mechanization is illustrated by the progressive amalgamation of task groups on a conventional cutting longwall when scraper-packing was introduced. Though difficulties in overcoming traditional attitudes prevented full amalgamation of all groups, the model is likely to be of general interest for the introduction of higher mechanization on conventionally organized longwalls. Cross-cultural evidence on the productivity gains obtained from composite organization on fully mechanized faces is provided by experiments carried out in the Donbass and this is compared with the experience of the research team.

EVALUATING THE MANAGEMENT OF CHANGE PROCESSES

To give a general orientation to the management of change processes, which is the underlying theme of the next two sections, it is necessary to anticipate some of the findings. Of the nine changes, seven were successful in that they eventually accomplished what they set out to achieve; the eighth was successful as a production unit by the standards of the pit concerned but failed to develop the intended form of composite organization; the ninth also was operationally successful, though at a lower level of mechanization than that attempted. The change process in two of the units was unduly extended in time; good results were accomplished only after very considerable difficulties—to some extent technological, but largely organizational.

At a somewhat different level of evaluation it may be asked how far the way in which the change process was managed enabled the inappropriate aspects of the existing work culture to be 'thrown off' and how far it permitted the emergence of optimum system characteristics. The studies indicate that in six of the situations there were identifiable limitations in the handling of the change. The need to follow the steps by which the working group developed an effective structure, to make use in planning of relevant recent experience, and to recognize those aspects of work organization requiring more than fortuitous dissemination was little appreciated. The major difficulties which arose centred on the socio-psychological system, to some extent because of preoccupation with the technological aspects, but largely because of a lack of concepts for analysing social process.

In conformity with the step by step approach, change was usually limited to one part of a seam or pit. As soon, however, as mechanization projects on anything but a small scale are contemplated the relation of haulage, winding, and surface arrangements to face capacity attains crucial importance. Not only does the total pit system become involved, but the use of Group and Area resources. This rapid expansion in the size of the *implicated system* is one of the main constraining forces inhibiting the introduction of change in ways likely to be maximally effective. The hypothesis is put forward that the successful implementation of change requires the exercise of continuously active leadership at the level of the largest directly implicated system. Whatever their point of initiation, the most successful change projects in the experience of the research team were those which had explicit

Area sanction and active Area support. Area, being the largest system directly implicated in productive operations, is the most effective change agent.

THE NEED FOR PROTECTION

The feature common to all these situations—though its degree varied in each case—was the tendency, described in Chapter XXII, to treat faces undergoing change as if they were ordinary new production units. In the experience of both management and workers they were operational experiments, not that anyone thought in terms of such a concept.

In their early phases, especially, operational experiments need *protection* from the requirements of production. Any protection afforded was all too often terminated as soon as the unit began to appear technically successful—or when it was felt that the men ought to have 'settled down'. Social adaptation, however, is a slower process than technological innovation. Too rapid removal of protection can precipitate social crises even when the new technology has become familiar. During a phase of protection the social as well as the technological system is accorded some attention, even though this may be no more than intuitive or peripheral. To this extent it is subject to control. With the termination, however, of explicit technological protection even rudimentary safeguards are removed and unrecognized changes in the social system may take place. Rice (1958) in the study parallel to the present research suggests that three phases of protection are necessary:

> '1. *Conceptual protection* (*complete*)—required in the design stage. The premature leakage of information about technological change that may have economic or social consequences may cause so much interference, both manifest and latent, that full exploration of the economic and social dimensions is inhibited. Models of a new system capable of dealing with the social and economic problems are never discovered.
>
> '2. *Experimental protection* (*substantial*)—required to allow the testing and modification of a new system under experimental conditions. Because any new system has eventually to become

part of a normal industrial unit, complete protection during testing may give rise to rumours, and so set up resistance in the unit for which it is intended. Against this, too little protection may result in too early crystallization in the social and economic dimensions, because of anxiety about the disturbance of traditional patterns.

'3. *Operational protection* (*limited*)—required during the transition from experimental testing to a normal industrial unit. The transition is a handing-over process, whose length varies with the deviation of the new from the traditional system. It may also vary according to the dimension in which the change is required. If little change is necessary in the social dimension, hand-over can usually be quick. Considerable social change may take longer to assimilate.'

Except in one case—that of the Haarman scraper-peeler—there were no indications that any of the change situations described were approached in this frame of reference.

One of the most striking findings was the lack of dissemination of the experience—however empirical and unconceptualized—gained in these operational experiments of patterns of work organization, not only among neighbouring pits, but even in the same pit and sometimes in the same seam. This did not arise because of any lack of facilities, but because the vigilance of colliery management in disseminating technological experience did not extend to the socio-psychological dimension. As yet, there was no establishment in the work culture of the idea of *social learning through operational experiment.*

The absence of this concept cannot but hinder the transition from the values and practices of unmechanized and partially mechanized mining to those more appropriate to fully mechanized continuous systems. Some of the more fundamental problems of change in work culture and their implications for operational management are examined in Section Eight.

CHAPTER XXIV

Change within Conventional Longwall Technology[1]

The change processes described in this chapter were observed during a study of conventional hewing longwalls in a pit where hewing was the main method of mining. They concern a change-over from single- to double-unit working and the subsequent introduction of cutting in one district of the seam in question—the Manley. These changes were the principal events in a pattern of development which brought the seam system from a steady state of low production to a new equilibrium at a higher level of productivity. Arrangements remained within the framework of conventional longwall working, though with an intensified composite component.

The seam had for many years been worked by single place methods and later by single-unit hewing longwalls. A tub haulage ran from the shaft bottom to the loading ends of the face belts, there being no gate or trunk conveyors. Two separate districts were being worked. The West with three single-unit longwalls was extremely wet, while the East with one single-unit longwall and a few single places had good conditions. The 90-yd longwalls were each manned by 12 hewers who worked in two shifts of six, and three pullers and eight stonemen who worked on a third shift. The hewing teams were cavilled each quarter, whereas the pullers and stonemen remained with the face for the whole of its life.

The manager, who had recently been appointed, found considerable concern at Area over the low productivity and high costs of these single units. The expected o.m.s. from the two hewing shifts was 7·5 tons, but the average hewer's o.m.s. for all

[1] Source paper: 16.

faces was no more than 5·7 tons—76 per cent of that expected. The manager believed improvement would follow from the concentration of working in each district by the introduction of double-unit faces with mothergate conveyors. Concentration of men, materials, and services at one place would permit better supervision, and the gate conveyor would eliminate hold-ups likely to occur when tubs were loaded directly from the face. Higher management was not optimistic about double-unit working because of an unsuccessful previous attempt; after faces had advanced 600 yds the mothergate had closed to such an extent that the gate belt had been barely able to run and the team had had to crawl through. Low output, high costs, and general dissatisfaction had resulted in a return to single-unit working.

TABLE 23 PRODUCTIVITY OF HEWING LONGWALLS IN TWO DISTRICTS BEFORE AND AFTER CONCENTRATION (HEWERS' O.M.S. ONLY)

Period	*East District*		*West District*	
1st Quarter (First half)	1 Single unit	7·1	3 Single units	5·2
1st Quarter (Second half)	1 Double unit (experimental)	4·6	2 Single units	5·7
2nd Quarter	„	6·4	„	4·7
3rd „	„	6·4	„	4·6
4th „	1 Double unit (operational)	8·3	„	4·0
5th „	„	8·0	1 Double unit	3·6

A DOUBLE-UNIT INNOVATION

After examining the difficulties, the manager decided that with improved supports double units were possible, provided they did not advance more than 150 yds. An experimental double unit with two 90-yd faces was started in the East district, manned by 24 hewers, six pullers, and nine stonemen. During what remained of the first quarter the hewers achieved an o.m.s. of 4·6 tons—less than that obtained from the remaining single units (*Table 23*). Such a difference during a period of experimentation was not unexpected. In the two subsequent quarters the hewers increased their o.m.s. to 6·4 tons, whereas that of the single units declined. The experimental double unit was now achieving 85 per cent of its expected output against 62 per cent for the single units. At the end of the third quarter, when the face had advanced 150 yds, the mothergate began to close. The experimental unit was stopped and another double unit, on an operational rather than an experimental basis, was started. The o.m.s. attained by the hewers was 8·3 tons—appreciably above the 7·5 tons expected for the conditions (112 per cent). In the meantime the o.m.s. on the two single units fell to 4·0 tons—53 per cent of that expected.

These differences reflect the worse conditions obtaining in the West district, where excessive water interfered with the running of face belts and was apt to hold up stonework. The declining output of the West, however, and the gains made by the East cannot be attributed to the physical differences, which remained constant. It is necessary to examine how the faces were manned.

The task of hewing and filling off the coal from both sides of the double unit was the collective responsibility of 24 men, six of whom manned each face on two successive shifts. The 24 men constituting the hewing task group shared equally in a common paynote, the size of which was determined by the weight of coal produced. When the double unit was being planned, there was widespread scepticism that so large a group could successfully share a common note. It was predicted that individual differences in performance would cause so much disagreement that the group would break up, or that it would retain its cohesion only by restricting its productivity to that of the slowest member—and, therefore, fail to 'make the price'. This prediction—which proved false—was based on beliefs current in single place working (Chapter IV). What it overlooked was the possibility that the older tradition might be constructively adapted to the new requirements. Such an adaptation took place in the team in a way which gave it a certain degree of compositeness. It will be shown in what follows that this contributed to the unusual degree of success of the panel.

The colliery had always been a three-shift pit so that three-man marrow groups were traditional. These groups had persisted in the longwall hewing cavils since a number of single places remained in the seam to which men might at some time go back. Groups of three who mutually accepted each other had formed up with similar groups to make teams of 12 hewers for the single units.

The experimental double unit began in the middle of a quarter when the single unit in the East district reached its boundary. The team from this face, together with one taken from a single unit in the West district, was moved in accordance with the cavilling rules on to the double unit until the end of the quarter. When the new quarter began the 58 hewers in the seam were called upon to marrow themselves into sets so that allocation of teams to the experimental double unit and the two single units could be decided. The two single unit teams placed by management on the experimental double unit had found during the six weeks of the quarter which remained that they had the same standards and worked well together. They therefore decided to amalgamate and to put in for the double unit as a combined set. They formed a kind of 'A' stream among the hewers in the seam, a set recognized by both management and the pullers and stonemen with whom they had worked as more able than the others. The teams on the two single units also decided to amalgamate—forming a 'B' stream—in order to compete for the double unit. To obtain the required number they had either to amalgamate or to break up their existing sets and attempt to form a new large group which would invite some of the men in the single places to join up with it. The single place groups were, however, all satisfied and did not want to change their arrangements. The cavils were put in first for the double unit and the A team was successful. The B team reverted to its two original 12-man units—which also formed 'a' and 'b' stream groups. These cavilled for the two single units. The remaining 10 hewers were cavilled in marrow groups of two or three to the single places and provided a reserve of reinforcements and substitutes for the longwalls.

A poor hewing team can substantially lower the earnings of pullers and stonemen, whose opportunities depend on the advance which the hewers can achieve. At each quarterly cavil the pullers and stonemen who stayed permanently with the double unit became anxious lest the B stream hewers should be their next partners. These pullers were the best in the seam and were eager to keep up with hewers who accepted a 6-yd advance during the week rather than the prescribed 5 yds and who achieved this for at least one side of the panel two-thirds of the time. This meant that the pullers got in an extra pull and that the stonemen could advance the gates further. Inferior hewers could not have kept up this pace.

On the double unit good hewing and pulling teams had come together neither by their own choice nor by management plan, but by the chance of cavilling. At the end of the second, third, and fourth quarters, the A stream continued to be successful in drawing the double unit so that it was able to establish a group standard to which all members both wished and were able to conform. The power of this norm was such that, when a man was absent from a shift group of six hewers on one of the faces, the others would still aim to reach the target. Substitutes were asked for only if it became evident that the target could not otherwise be reached. As the group was an elite, the norm established was at a high standard, the level of output being 12 per cent above that expected. As well as being manned by the less able stream, the single units had to contend with bad conditions. Their repeated lack of success at the quarterly cavils lowered their morale and with it their output. While the 'good' piled up in one district, the 'bad' piled up in the other. Cavilling had produced the type of result it was instituted to avoid.

AN ATTEMPT AT EXTENSION

At the end of the fourth quarter one of the single units—manned at that time by the more able of the single unit teams—reached its boundary. The manager, in pursuance of his policy of concentration, opened up a 65-yd face on the other side of the mothergate occupied by the remaining single unit and installed a gate conveyor. Although the older 90-yd face was 30 yds ahead of the newer 65-yd face, the layout was treated as a double unit and the two face groups were amalgamated. The two hewing teams retained their identity by each taking over a whole shift on both faces. This was the opposite of the cross-shift organization of component marrow groups which had emerged on the original double unit (*vide* p. 229). It was not successful. At various times there were demands that the two hewing shifts should be paid on separate notes. The two shifts, however, are not equivalent since the first, which has to break in, has the harder job. The teams could not agree to cavil for the harder shift and had no option but to remain on the same note. Tension had developed because one group was made up of older and more skilled men, while the other contained the younger and less skilled. The conflict was displaced on to management who were perceived as attempting to undermine the manning clauses of the agreement 'because they were coal hungry'. Though one face was 30 yds longer than the other, the men insisted on keeping six hewers on each, holding that this difference was wholly neutralized by the harder coal and wetter conditions on the shorter face. When team members were absent they kept six men on the longer face irrespective of the state of the cycle. Tension increased as the coal hardened and on one occasion it took five hewing shifts to get a 2-yd advance over the panel as a whole so that the earnings of pullers and stonemen suffered severely as well as the hewers' own. The standard 5 yds per week was never achieved for both faces and only once for either. On one shift, when coal had been left on, three pullers removed more than the whole hewing team had managed on the previous shift.

The tension within the hewing task group on this second double unit may be attributed to the fact that its two constituent shift groups were of different capacity—a situation which arose because they were distinct groups in each of which men of similar ability had chosen each other. Each of these constituent groups maintained its separate identity by taking over a whole shift. On the first panel group boundaries did not develop exclusively on shift lines. The set was composed of three-man marrow groups, and one member of each always placed himself on the other hewing shift. This practice had two functions: it kept a place open at the alternative shift time should members require a change from their ordinary schedules; and it provided a representative of each group's interests on the other shift. Accusations of one hewing shift by the other never passed beyond the stage of good-humoured joking, whereas on the other unit they were a major source of ill-feeling.

RECONSIDERATION AND TECHNICAL CHANGE

At the end of the fifth quarter the two hewing groups on the second double unit could not agree sufficiently to make up a 24-man set to enter the cavil. Having previously worked on different single units, they had had no actual experience of each other's standards until the fourth quarter, whatever they may have believed these to be. Nevertheless, once the group on the first double unit had decided to stick together, the single unit teams had no option but to amalgamate if they wished to compete for the panel with better conditions in the East district. They had been consistently unsuccessful at the cavilling and only experienced working together when the second double unit started. Differences in capacity and skill, undoubtedly exacerbated by the bad conditions, were too great to permit co-operation.

The teams forming the first double unit had been more favourably placed: having been brought together experimentally, they found themselves compatible and *chose* to amalgamate. Had they not accepted each other, they still had the freedom to join up with either of the other two single units and to take a chance on compatibility working out in practice. The groups forming the second double unit had no choice. They were *forced* to amalgamate, whether compatible or not. As they were not, they faced the dilemma of continuing a union which was not working out, or suffering dispersal with possible loss of status and earnings.

During the fifth quarter hewers' o.m.s. on the second double unit (3·6 tons) was less than half that on the first (8·0 tons). Results had become so poor—because of bad conditions, increasing hardness of the coal, and poor capability and lack of cohesion in the team—that the manager decided to stop the shorter face and extend the longer as a single unit, at the same time changing over to a cutting technology. He expected to get more coal by using a 4′ 6″ cutting jib on a single face of 100 yds than he had obtained from both sides of the hewing panel.

Just as there had been beliefs that the seam was unsuitable for double units because of the roof, so there were beliefs that it was unsuitable for cutting because of the floor. None the less, improved technique made cutting possible. A team of nine fillers was required, and two sets—one from each of the hewing teams—applied. The cavilling resulted in a win for the younger and less experienced group, the older and more experienced being dispersed throughout the pit to single places and bargains. Despite this apparent setback, production increased and five cycles per week were regularly completed. At the same time the original double-unit hewing longwall maintained its productivity. As a side effect of the change-over to cutting in the other district, the cavilling problem was solved, for hewers could not according to custom be cavilled with fillers, with the result that each group remained with its own face.

Before the introduction of the double unit the seam was, with its various single units, in a steady state of low production. The advent of the double unit disturbed this condition and, though beneficial in itself, created a situation in the two remaining single units to which there was no effective solution within the cavilling rules. The replacement of a hewing by a cutting longwall brought the seam system into a new equilibrium at a higher technological level, and disposed of the cavilling problem. The face population was no longer split into an idealized group in a 'have' and a denigrated group in a 'have not' position. A situation existed which permitted working districts to consolidate the relations of their various task groups, all of which now had equal investments in their face 'enterprises', to which they were committed so long as these lasted.

CONCLUSIONS

This course of events demonstrates the need to select the appropriate system when change is introduced. Changes internal to face systems, even though well within the limits of conventional longwall working, may have wider effects which can be assimilated only when the seam as a whole is reorganized. In the East and West districts the success of the changed faces highlighted the failure of the unchanged, inducing the need for further change. When, however, a second change was introduced, as so frequently happens in such circumstances, it was modelled directly on the first. Given the cavilling rules, this could only produce a stalemate, which persisted until a change in the physical environment—the hardening of the coal—provided the rationale for a purely technological change—the introduction of cutting—which had the side effect of solving the unlooked-for problem that had arisen in the socio-psychological system.

To proceed one step at a time and to try again what has previously succeeded is the most commonly adopted approach to the practical management of change situations. Though having the merit of cautious empiricism in entering complex situations, where the unknowns are many, it has little power to detect the latent effect of local changes on wider systems. In the present case, a more systematic approach might have attained the new steady state and realized the full production gains in substantially less time than the year and a quarter actually taken. Such an approach, however, would have entailed a much more comprehensive initial analysis than any likely to have been made in the prevailing 'ethos' of the management climate. Action would have been necessary at seam rather than at face level and in terms which took into account socio-psychological as well as technological factors. Any such course would have enlarged the scale and altered the pattern of what was being done in a way that would have placed it outside the norm of the type of change common in the environment of this colliery. This norm was sanctioned by the common expectations of higher management, pit officials, workmen, and trade union representatives. To break out would have led to many further complications since a new force field would have been entered. Yet only by breaking out could better results have been obtained.

In terms of the step by step theory of change introduction, the manager of this colliery had given what would be regarded as a model performance. He had applied sound mining theory which had been proved right in the face of disbelief; on the other hand, he had not been over-ambitious and had accepted a good deal of trial and error. He had also had the good sense to leave the cavilling problem to resolve itself in the course of time and as opportunity presented. By proceeding in this way he had carried everybody with him and had effected a worthwhile improvement without incurring labour trouble. It is only in a quite different frame of reference that the limitation of such an approach can be seen and the question asked as to what alternative concepts and methods might have brought about the same amount of change in, say, six rather than fifteen months.

This study also provides examples of both appropriate and inappropriate persistence in longwall organization of features of the single place tradition. Cavilling proved maladaptive. By contrast, the cross-shift organization of the component marrow groups in the hewing double unit was the key to the unexpected success of the large identical role group involved.

Though a single task group, and in this sense conventional, this team behaved as a composite work group (Chapter IX). It practised task continuity between the breaking in and the later phase of hewing so that no friction was created between the two shifts. There was no 'made' work and the face was kept clear of spillage. Timber was meticulously in alignment, which distributed weight on the face in a way that facilitated hewing activities and so made additional advance more attainable. Men deployed themselves flexibly on a panel-wide basis to prevent lag developing in particular sections or to win an extra advance which would permit a double pull. There was a high degree of self-regulation and, since the group (despite absentees) attempted to reach the production target without external reinforcement, a high degree of self-maintenance. Through the cross-shift marrow groups, interchange of shifts was possible so that individual needs were met and stress reduced. As the principle of equivalent contribution was recognized, the total team of 24 was content to share a common paynote.

The record of this team shows the social and psychological resources at the disposal of men with a single place background in using composite principles for the creation of a relatively large work group. From the successful self-regulation on these principles of the largest single task group in conventional longwall working —the two-shift, panel-wide, hewing team—it would seem but a step to include the other and smaller task groups in an overall cycle organization. Nevertheless, this step is a very big one, and had not been taken in the pit discussed despite its closeness to the single place tradition.

Once hewers have found a way of regulating themselves as a total group it might be expected that barriers to complete composite organization would be fewer and more easily surmountable on hewing than on cutting faces. There are only two other types of group concerned and differences in status and earnings are less between pullers and hewers than between pullers and fillers. Moreover, since hewing occupies both the fore- and the backshift, pullers and stonemen are condemned to permanent night work, and might be presumed not unwilling to grasp an opportunity to change this situation. In the present case unrelieved nightshift was accepted fatalistically by those concerned, who conceived of no other possibility. Yet the colliery in which composite longwalls had developed under pressure from men who objected to being permanently on bad shifts was their close neighbour, whose winding gear on a hill on the other side of the valley could be seen from the pit yard. No information, however, regarding the changes in work organization which had recently taken place there had crossed the valley. The grapevine remained silent, while inter-pit trade union meetings stuck to wages and conditions, and management conferences to technology and costs—or labour relations. This lack of communication is scarcely surprising when so little information about the new type of work organization had passed from one seam to another in the pit originally concerned or, indeed, between adjacent faces in the same seam operating the same agreement. Work organization had not yet been recognized as a dimension of either formal or informal discussion. Concepts were lacking to identify the problems and clarify the issues.

However much developments may have stopped short in this way, the research team regarded this pit, because of its capacity to adapt the single place tradition, as particularly fertile ground for the introduction of composite longwall working. Single task mechanization was planned for the Manley seam and a rewarding field experiment would have been to follow equivalent sets on to the new faces, had one elected to proceed in single task groups and the other as a composite team. The research period, however, ended before such a development took place.

CHAPTER XXV

Changing from Conventional to Composite Working: The Process[1]

THE SEAM BEFORE REORGANIZATION

The Background

The successful introduction of composite organization for the three-shift cutting longwall cycle in the Manley (*vide* Section Five) encouraged management and lodge in the early months of 1955 to consider introducing the same type of agreement in other seams. In each case the period of negotiations was shorter than that preceding the initial agreement, yet sufficiently extended to draw attention to new issues.

When in July 1955 visits were paid to conventionally organized faces in various seams, including the Bramwell, it was noted that, should a composite agreement be concluded here, a crucial test of the viability of the composite form of longwall organization would be provided. The most striking features were the extreme diversity of working conditions ('good' and 'bad' cavils), the existence of three different technologies (hewing and cutting longwalls together with single places), and two different forms of work organization (conventional and composite). Further sources of variation arose in differences of face length (33–110 yds); in gateway stonework which in some cases was a 'second ripping' up an old gate; in face and gate conveyors (rubber belts, Blackett and P.F. scraper chains); in roof supports (three types of chock); and in travelling time between bank and kist, etc.

The coal was 26″–28″ in height though weekly averages ranging between 21″ and 34″ were recorded. It was free from band, but

[1] Source paper: 14.

some areas were subject to minor rolls and hitches, and one to major faulting. On the south side extremely wet working was encountered.

Before reorganization the main production had come from four single-unit longwalls—one organized as a hewing face on conventional lines, one as a conventional cutting unit, and two as cutting units on the then existing composite agreement for short-walls. In addition, on a small cutting panel the faces were worked alternately, but, as the team members were used as substitutes on other faces, its production was irregular (face S). There were eight three-man hewing sets who worked in the single place tradition winning new faces and clearing out small areas of coal, and three small groups of bargain stonemen similarly employed. *Table 24* summarizes the technological, environmental, and organizational characteristics of the principal production units.

The size and composition of the longwall face teams varied between 14 and 28 men. On the two composite faces the workers had marrowed themselves as one group and had been together from the beginning. The three conventional teams were separate task groups with fillers and hewers cavilled every quarter. Though the teams were constituted according to the agreements for each type of working, the level of interference arising from geological causes or haulage disturbances produced such a chronic state of dysfunction that men were deployed to reinforce one team at the expense of another. The achievement of five cycles per week was exceptional (a chance of one in twelve) and each face had a different expectancy—varying between two and four cycles per week. The formal allocation of men to shifts on the conventional faces was often at variance with the requirements of an out-of-phase cycle, in terms sometimes of skills and sometimes of numbers. Under such conditions a mobile reserve was necessary, but because of absences this reserve was inadequate. The multi-skilled composite teams were located in a remote part of the seam. Though they could reinforce each other, neither was near enough to reinforce the conventional faces. In consequence, many of the men on the conventional faces acquired a broader experience of facework than their formal qualifications and assigned work roles might suggest. Indeed, at times what amounted to task continuity was practised.

TABLE 24 CHARACTERISTICS OF MAIN PRODUCTION FACES DURING QUARTER BEFORE REORGANIZATION

Face	*Work Organization*	*Technology*	*Face Conveyor*	*Face Length and Type*	*Conditions*	*Team Size*	*Work Roles*	*Mean Cycles per Week*	*E.M.S.*[1]
A	Conventional	Cutting	Blackett scraper	Single unit 85 yds Bordways	Dry Bad faults	24	2 cutters 1 driller 10 fillers 5 pullers 6 stonemen	4	106 (fillers)
B	Conventional	Hewing	Rubber belt	Single unit 110 yds Bordways	Wet Minor hitches	28	18 hewers 4 pullers 6 stonemen	3	100 (hewers)
C	Composite	Cutting	Blackett scraper	Single unit 90 yds Headways	Dry	18	Composite	4	144 (all)
D	Composite	Cutting	P. F. scraper chain	Single unit 48 yds Headways	Very wet Hitches	23	Composite	2	103 (all)
S	Conventional	Cutting	P. F. scraper chain	Double unit 33 yds each face Headways	Wet	14	2 cutters 4 fillers 2 pullers 6 stonemen	2	102 (fillers)

[1] E.M.S. = Earnings per Manshift; lowest = 100.

Because of the low level of cycle completion, strict implementation of the agreements would have placed some faces more or less permanently on the minimum wage. Recourse to making-up resulted in a steady level of wages on each face, earnings and production bearing little relation to each other as may be seen from the last column of *Table 24*. The average earnings per manshift were 15 per cent lower than on the composite longwalls in the Manley.

Following the conclusion of the composite agreement, visits were paid to the seam on eight occasions from January to April 1956. Discussions were held with management and lodge officials and a study made of seam records for three months before and eight months after reorganization.

Attitudes towards Composite Working

Despite the fact that they were members of the same lodge and lived in the same village, the men had only the haziest idea of the system of composite longwall working developed in the Manley. They attributed its success to better conditions, good workmen, and a higher wage level. Almost nothing was known about the task-shift rotation systems and the kinds of work role which had evolved, and very little about the basis of payment and the equal sharing of the common note. What had happened in the Manley was regarded with indifference, partly because of the relative independence of the seams and partly because neither management nor lodge had taken active steps to communicate the available experience.

Few men on the conventional faces expressed themselves as in favour of composite work. The majority preferred 'one man—one task' assignments. What they disliked was 'chopping and changing jobs during a shift' and 'not knowing what job you would be doing when you came on'. They were referring to the shortwalls, which were their only sources of information, and their dislike was reinforced by the quasi-composite conditions induced on conventional faces by cycle dysfunction.

REORGANIZATION AND ITS EFFECT ON TEAM COMPOSITION

The Course of Negotiations

Such was the situation when management and lodge opened negotiations during the spring of 1955 with the Manley agreement as a model. The main issue centred on the diversity of working conditions. Management wanted one agreement to cover all conditions. The lodge considered this unfair. During July and August further progress was interrupted by a dispute, but at the end of September the lodge called a meeting to consider prices and manpower. A notice calling on the men to marrow themselves into sets was posted on 8th November, the lists to close at 11 a.m. on the 11th, but because of confusion over the cavilling rules the closing date was put back to the 18th. Even then prices had not been settled and a meeting with higher management was necessary before an agreement was signed on 21st November. Sets were then cavilled to faces in the traditional order—'sun ways round'—and composite longwall working began in the Bramwell on 28th November.

The misunderstanding over cavilling rules arose because a group who had marrowed themselves into a set of 27 (the largest required) for one of the better faces believed they would get the face they wanted if no other set of 27 was formed. Their view was opposed by men in difficulties over completing so large a set, since it implied their restriction to the poorer faces. The lodge gave the following ruling: sets of 27 should be organized for all faces, the men nominating a number of reserves who would fall out should the set be cavilled to a shorter face. These reserves were men qualified only in filling and stonework, for cutters and pullers were at a premium.

The Initial Reorganization of Face Teams

Reorganization meant a one-sixth reduction in manpower. Some of those made redundant transferred to facework in other seams, others preferred to become datal workers. A few continued as substitutes, this being feasible because teams kept open the places of men who, through illness or injury, were absent or employed on 'light work'. Similar provision was made for marrows qualified and employed as spare shotfirers or deputies.

The extent to which those remaining were regrouped may be seen from *Table 25*. Only one set emerged intact—a composite team which moved from Face *C* to *D*. The other composite team was split and the collections of single task groups on the conventional faces were completely broken up. Existing marrow groups however—whether task groups from conventional faces, single place sets, or larger groups from composite faces—were the 'bricks' from which the new teams were built. The outcome was that the two largest teams, of 27 men each, were composed of men from nine and ten different task groups; two of the shorter faces had men from four and six groups; only in team *D* did all the men come from the same group.

Subsequent Face Reorganizations

By the time the reorganization was implemented three of the existing faces were nearing their boundaries and three new faces (*E*, *F*, and *G*) had been prepared of somewhat different dimensions from the old faces and requiring teams of slightly different size (25, 24, and 24 respectively). After only three weeks *A* finished and its team was to move to *E*. However, the team on *S* laid claim to *E* on the grounds that management had referred to *S* as a 'spare cavil', hence the rule of 'first out—first in' should apply. The issue was discussed between management and lodge and, though the designation had been unwitting, the cavilling rules were upheld. The 17 men from *S* moved to *E*, the set being made up to 25 by cavilling eight men from the *A* team, the remainder of whom went into the spare cavil as team S_2. This team was in fact 17, not 19, as team *A* had been working with two men short.

This incident established a precedent for all subsequent changes, the team currently occupying face *S* having first claim on any new face (*Figure 18*). Eight weeks later *B*, which was on the North side of a mothergate, finished and *F*, which was on the other side of the same gate, was ready for a team of 24 men. The *B* team considered disputing the cavilling rules on the grounds that *F*

TABLE 25 ORIGIN OF NEW COMPOSITE FACE TEAMS

Old Face	*Task Group*	*New Face Teams* A	B	C	D	S	*Men not on new faces*
A (Conventional)	Cutters			2			—
	Driller			1			—
	Fillers	1		9			—
	Stonemen (M.G.)[1]	2					2
	Stonemen (T.G.)[1]		2				—
B (Conventional)	Hewers	6	7				5
	Stonemen (M.G.)	4					—
C (Composite)	Composite workers				18		—
D (Composite)	Composite workers			6		14	3
S (Conventional)	Cutters		2				—
	Fillers			2		1	1
	Pullers	1					1
	Stonemen	2					—
Single Places (Coal Bargains)	Hewers a[2]	3					—
	Hewers a	3					—
	Hewers b		3				—
	Hewers b		3				—
	Hewers b		3				—
	Hewers			2		1	—
(Stone Bargains)	Stonemen a	4					—
	Stonemen b		2				—
	Stonemen d		4				—
	Others	1	1			1	
Size of Team		27	27	22	18	17	—
No. of constituent groups		10	9	6	1	4	—

[1] M.G. = mothergate. T.G. = tailgate.
[2] The lower case letters identify the conveyor face closest to the single place.

could be regarded as part of their cavil, but nothing came of it and the 20 men not cavilled to *F* moved into the spare cavil as team S_3. A fortnight later *C* finished and its team split up between *G* and S_4.

Three weeks after work started on *E* a major fault was encountered and the face had to be rewon. This took 20 weeks, so that close on six months had passed before the seam as a whole became settled after the initial reorganization.

FIGURE 18

REGROUPING AND SUBSEQUENT RE-CAVILLING OF BRAMWELL FACE TEAMS

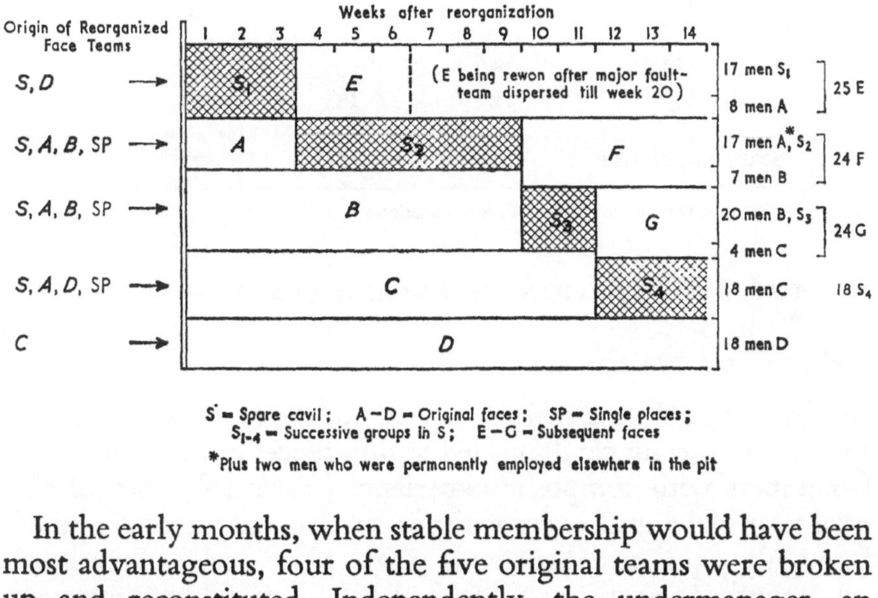

S – Spare cavil; A–D – Original faces; SP – Single places; S_{1-4} – Successive groups in S; E–G – Subsequent faces

*Plus two men who were permanently employed elsewhere in the pit

In the early months, when stable membership would have been most advantageous, four of the five original teams were broken up and reconstituted. Independently, the undermanager, an overman, and a faceworker used the same words to describe the consequences: 'It breaks up teams that have worked together and doesn't allow new teams to build up in a balanced way.' Had *S* not been designated as a spare cavil, no particular difficulty would have arisen from the small differences in team size between the old and new faces, since, even during the early weeks, some men withdrew from facework and were replaced by others held as spares. Permanent teams would have been established.

TABLE 26 PERCENTAGE DISTRIBUTION IN REORGANIZED TEAMS OF THREE TYPES OF FACE EXPERIENCE

Face Team After Reorganization	*Before Reorganization*		
	Composite	*Single Place*	*Conventional*
A	—	41	59
B	—	59	41
C	27	9	64
D*	100	—	—
S_1	82	12	6
E*	56	12	32
S_2	—	53	47
F*	—	54	46
S_3	—	60	40
G*	8	50	42
S_4*	22	11	67
Seam Population	34	28	38

* Faces in production at end of reorganization.

THE DISTRIBUTION OF EXPERIENCE AND SKILL

Previous Composite Experience

The element of chance which entered into the composition of the teams at the various cavillings led to differences in the proportion of members with composite experience (*Table 26*). Two of the initial teams (*A* and *B*) were without anyone and were composed of roughly equal numbers with single place and conventional backgrounds. Later, teams S_2, S_3, and *F* were similarly composed. The absence of men with composite experience meant that these teams had no member who could give guidance on the most appropriate ways of working and organizing under their new composite agreements. Of more immediate importance was the absence of multi-skilled men to occupy key roles. Where men with recent experience were not available, the teams were obliged to draw on those who, though formally qualified, had not practised a particular skill for some time. Apart from cutting, the ratio in all teams of those qualified to those required was greater than 2·0, which indicates scope for at least one change of occupant in each role. In cutting, the situation was much less satisfactory; three teams—*A*, S_2, and S_3—emerged from the cavillings without any qualified cuttermen, so that management had to supply them. Fortunately, none of these teams remained together for long. Two teams—*B* and *F*—had no reserve in case of absence, and no spare cutterman to work on the filling shift should the cycle lag. In contrast, teams *C*, *D*, S_1, and *E* were lavishly supplied, containing many composite workmen.

Previous Skill

The resources of skill in the seam were not inconsiderable (*Table 27*). Virtually all men were qualified in filling and stonework, one-half or more in pulling and drilling, and one-quarter in cutting.

TABLE 27 SKILL UTILIZATION IN THE BRAMWELL BEFORE REORGANIZATION

	Skill[1]					*Total*
	C	D	F	P	S	*Skills*
Percentage of all faceworkers qualified in each skill:	25	52	97	62	99	67
Percentage of qualified faceworkers employed on:						
Composite faces	79[2]	62	35	52	35	46
Other faces, using skill	17	2	43	2	18	19
Other faces, *not* using skill	4	36	22	46	47	35

[1] C = Cutting, D = Drilling, F = Filling, P = Pulling, S = Stonework.
[2] This and subsequent figures are percentages of 25, 52, etc.

Before reorganization only men on the composite faces were in a position spontaneously to make use of more than one of their skills—46 per cent of the resources available. On the conventional units separate notes and segregated task groups precluded the use of skills other than those which the particular agreements specified; only 19 per cent of the available skills were utilized; 35 per cent were dormant and could only be used by special arrangement.

With the introduction of composite agreements, the whole of the skill resources became potentially available. How far advantage could be taken of this was dependent on the extent to which task continuity was carried out and on whether appropriate methods of task and shift rotation were developed, i.e. on the degree of compositeness of the systems which developed.

DEGREE OF COMPOSITENESS OF VARIOUS FACE TEAMS[1]

Of the eleven face teams which at some time existed, four were in the spare cavil, *S*, and rarely functioned as a whole team; one, *A*, lasted for only three weeks before being re-cavilled; and another, *E*, was dispersed for 20 weeks while its face was rewon. Of the five remaining teams, two, *B* and *C*, existed for nine and eleven weeks respectively during the early weeks of the reorganization and two others, *F* and *G*, came into existence after all men had some experience of composite work. The fifth team, *D*, was quite different; it had worked as a composite team for many years and was cavilled from a good to a bad face—an event not unique in its experience.

Because face *E* was being rewon, an unexpected reserve existed. When faces encountered difficulties additional men were available and teams did not have to adjust their organization to cope unaided. This situation disposed to the perpetuation of conventional practices for handling cycle disorganization, and the emergence of organizational forms more appropriate to the conditions was to some extent inhibited.

In Appendix II detailed accounts are given of the outcome on *B* and *C*, *F* and *G*, and *D*. In order, however, to permit systematic comparison and to relate degree of compositeness to level of performance, a scale of compositeness has been devised on the basis of the four aspects of composite working described in Chapter IX. Where the situation in relation to a given aspect is

[1] After reorganization, the task structure of the longwall cycle in the Bramwell was somewhat simpler than in the Manley. All faces were single units so that problems of co-ordination between two faces did not arise. In addition, the mothergate was not carried ahead of the face, the cutting machine moving in an arc across the gate, where coal was filled off by the fillers so that there were no hewing roles. The cycle was otherwise as in the Manley.

TABLE 28 A SCALE OF COMPOSITENESS

Aspect	*LEVEL OF COMPOSITE PRACTICE*			
	0	1	2	3
Task continuity	Own task	Simpler forms of advance and lag control: cutters track through for fillers, pullers take left coal; caunch work as required	In addition to (1): belt moved over by fillers and cutter turned by stonemen	In addition to (1) & (2): cutting completed by fillers or started by stonemen; chocks started by fillers, completed by stonemen or cutters
Multiple roles and skills	One task	Interchangeability between scuffling and filling, filling and stonework	In addition to (1): at least one filler and one stoneman qualified to pull and another to cut	In addition to (1) & (2): sufficient reserve able to pull and cut to meet all contingencies
Task-shift rotation	One task on one shift, or on two shifts if filling times alternate	In addition to (0): special arrangements between individuals or groups not involving whole team	Regular traditional informal exchange for team as a whole (shortwall model)	Regular planned formal exchange for team as a whole (Manley model)
Payment groups	Equal to number of single task groups	Some fusion of single task groups	Two groups—e.g. face team and gate team	One inclusive group

TABLE 29 DEGREE OF COMPOSITENESS (BRAMWELL FACES)

Face	*Ratings*				*Total*
	Task continuity	*Multiple roles and skills*	*Task-shift rotation*	*Payment groups*	
F	1	—	—	3	4
B	2	—	1	3	6
G	3	2	1	3	9
C	3	3	2	3	11
D	3	3	3	3	12
Total	12	8	7	15	42
Possible	15	15	15	15	60

the same as on conventional longwalls a zero rating has been given; thereafter ratings of 1–3 are assigned according to the extent of composite practice. Since all faces shared a common note each carries a rating of 3 as regards method of payment. Theoretically, ratings of less than 3 are possible on this aspect as much as on the others, and a scheme for the progressive amalgamation of single task groups is given in Appendix III; the hewing group described in Chapter XXIV would carry a rating of 1 for payment and a further rating of 1 for the task continuity practised between the two shifts. The definitions for the values on the rating scales for each of the four aspects are presented in *Table 28*, and the ratings themselves in *Table 29*.

The degree of compositeness of the different faces varied widely, very much in accordance with the distribution of previous experience and skill. As time went on compositeness increased in the teams with lower ratings. Of the two Manley panels described in earlier chapters, No. 2, like *D*, carries a rating of 12, No. 1 (East) a rating of 9 during phase 1 and of 10 during later phases—and No. 1 (West) a rating of 10 throughout.

CHAPTER XXVI

Changing from Conventional to Composite Working: The Results[1]

PERFORMANCE RECORD

'Before' and 'After' Comparisons

If the same teams had remained on the same faces after reorganization without technological change, a straightforward indication could have been obtained of the effects of introducing composite longwall working, as for the Donbass faces described in the Report of the Technical Mission (Chapter XXX). With the regrouping of teams, however, the finishing of *A* and the introduction of cutting on *B*, only limited 'before' and 'after' comparisons are possible. Data were available for 12 weeks before reorganization and for nine, 11, and eight weeks after, on *B*, *C*, and *D*. The comparisons for each face (*Table 30*) are made in terms of the completeness of production index described in Chapter XII. As actual manning was the same in the before and after periods, it does not have to be taken into account as it has to be when comparisons between faces are made.

Face *B*, which changed from a conventional hewing to a composite cutting organization, showed a statistically significant 12 per cent improvement equivalent to an increase of two cycles per month. This is due in substantial part to the change-over to cutting, for the coal was said to be hard to hew and the new team, which functioned only during the first nine weeks after reorganization, had a comparatively low composite rating (6).

Originally manned by an established composite team of high repute, *C* was taken over by a newly organized group, one-third of whose members had previous composite experience. The fall

[1] Source paper: 14.

in completeness of 6 per cent does not, however, reach statistical significance ($p > \cdot 05$).

TABLE 30 COMPLETENESS INDEX (PER CENT) BEFORE AND AFTER REORGANIZATION

Period	*Face* B	C	D
Before Reorganization	64	81	30
After Reorganization	76	75	42
Difference	+12	−6	+12

On this face a conventional majority conformed to a composite minority able to place men with cutting experience on all shifts, while ensuring that an unusually arduous pulling task never lacked an expert approach (*vide* Appendix II). The team had a high composite rating (11) and it is suggested that losses would, in all likelihood, have been greater had there been no composite nucleus.

Face *D* remained as a composite cutting unit, but the team taking over (the high ranking group which had left *C*) had worked together many more years than had the team occupying the face before reorganization. The 12 per cent improvement (statistically significant at $p = \cdot 05$) suggests that the more experienced team was better able to cope with the adverse conditions, particularly as they had the more cumbersome Blackett in place of a P.F. conveyor during the period of the comparison.

Manpower Comparisons

The index of completeness gives the level of production relative to what is possible with the given technology, without taking into account the amount of labour. In any week total labour is the number of normal shifts, plus overtime—whether by team members, substitutes, or reinforcements—less time spent on shift work and waiting-on. The resulting figure expressed as a percentage of the scheduled number of shifts is the *manpower index* (cf. Chapter XII). Values greater than 100 indicate reinforcement, those less than 100 show that the team worked short-handed (*Table 31*).

TABLE 31 MEAN COMPLETENESS AND MANPOWER INDICES AFTER REORGANIZATION

Face	*Duration in weeks*	*Completeness %*	*Manpower %*
B	9	76	103
C	11	75	98
D (1)[1]	8	42	74
F	25	77	112
G	25	95	103
D (2)[2]	26	51	72

[1] with Blackett face conveyor (P.F. conveyor before reorganization).
[2] with bottom loading belt conveyor.

Management's deployment of available manpower in order to maximize production from the seam as a whole means that the completeness index reflects the ability of the team to cope with the conditions on the face only when the scheduled manpower was employed.[1] Reinforcement above the scheduled figure may be necessary because of geological difficulties or because none of the team members on a particular shift has the required skills. The size of teams was agreed by management and lodge as appropriate to the dimensions of the faces and permanent features of the conditions. Abnormal geological conditions such as faults or 'freak' coal heights were not a permanent feature, but infrequent and of relatively short duration. If teams were functioning effectively, it was to be expected that, over a period of months, the average manpower index would be in the neighbourhood of 100. The amount of permanent reinforcement indicates the extent to which the team has been unable to cope with the general run of conditions.

[1] Faces *B*, *C*, and *D* were simultaneously in operation. The differences in manning reflect a management policy of diverting available substitutes and reinforcements to the two longer faces (*B* and *C*) which were the biggest producers. When *B* and *C* finished they were replaced by *F* and *G*. On *G* with its good working conditions manpower was maintained slightly above the scheduled level, while on *F* considerable reinforcement was given to maintain production at much the same level as on the face it had replaced (*B*). The policy of not maintaining *D* was continued.

Change Processes

Faces *F* and *G* had an average manpower index in excess of 100 over a period of six months—112 for *F* and 103 for *G*. Face *G*, composite rating (9), reinforced only to the extent of 3–4 shifts a week, had a balanced team: all operations could be carried out on any shift, this being particularly important in the case of cutting. The bulk of the men had worked together on previous faces; there was a nucleus with previous composite experience and, although single task work roles were developed, some interchange occurred.

Face *F*, composite rating (4), was reinforced on the average by 3–4 men per cycle or 12–16 shifts on a four-cycle week. The team was unbalanced; and, as in *B*, special men had to be brought in to complete cutting on the filling or pulling shifts. The team had not worked together before and none had composite experience. Work roles were rigidly of the single task type, and because of a lack of other skills among the fillers, who were shift rather than cycle oriented, the full possibilities of task continuity were not realized. With growing experience, however, less reinforcement was needed—in the first three months the manpower index was 114, in the second three months, 110.

Face *D*, composite rating (12), continued under-manned. Substitutes for absentees were rarely sent in, reinforcements never, and one-sixth of the shifts attended by the team were spent on shift work away from the face or as reinforcements to other teams. In view of the difficult working conditions, management expected lower production—two or three cycles each week—and did not attempt to maintain the team at full strength. Nevertheless an improvement was obtained when the Blackett conveyor was replaced by a bottom loading rubber belt.

Completeness/Manpower Ratios

On the hypothesis formulated in Section Three it would be expected that the overall effectiveness of the Bramwell faces would correspond to their degree of compositeness. In broad terms this may be shown to be the case. If the wet face *D* is omitted, the geological conditions on the other four faces are approximately equivalent, and technological equivalence may be obtained by making a correction on *C* for the Blackett conveyor. This may be estimated from the improvement from 42 per cent to 51 per cent in completeness obtained on *D* when a bottom loading rubber belt was introduced. On the assumption that a proportionate improvement would have occurred on *C* from similar technological change, a completeness/manpower ratio of 93 is obtained.

TABLE 32 COMPLETENESS/MANPOWER RATIOS IN RELATION TO DEGREE OF COMPOSITENESS

Face	*Completeness* / *Manpower*	*Composite Rating*
C	93	11
G	92	9
B	74	6
F	68	4

The completeness/manpower ratios and composite ratings are in correspondence. The striking feature is the size of the difference between the two quasi-conventional faces (*B* and *F*)—where, apart from the method of payment, the features of composite working had scarcely come into existence—and the two faces with higher composite ratings (*C* and *G*)—where task continuity was fully practised. This result is in conformity with the general theory of composite working elaborated in Chapter IX where first importance as regards effect on productivity was given to task continuity. The differences are less than those obtained from the comparison of extremes made in Chapter XIII but more than those between the faces with high and closely similar composite ratings compared in Chapter XIV. All results are in the expected direction and are self-consistent.

It was also suggested from the comparison made in Chapter XIV that the full development of multi-skilled roles and task rotation systems is likely to affect the level of stress and the ability to maintain the steady state under adverse conditions. The Bramwell data also support this view—as in the improved completeness on *D* when the more experienced composite team took over.

CONCLUSIONS

The gross result of organizational and technological change in the seam was a considerable improvement, the three main production faces together averaging eleven cycles per week compared with nine cycles before reorganization, an increase of over 20 per cent (cf. Donbass data, Chapter XXX). Some of this was due to the change on one face from hewing to cutting and to better face and trunk conveyors, but the main change was organizational. The new composite agreement made available dormant skills which the old system of separate notes had failed to use. The degree of improvement increased as time went on—and new adaptation, relationships, and learning were consolidated.

Yet adherence to traditional methods of forming and allocating teams resulted in sub-optimum deployment of the experience now released. The initial concentration of men with composite backgrounds and of qualified cutters and pullers in some teams produced others that were deficient in members with these skills. The marrow tradition of self-selection cannot give the best results in a new situation if no directives are given which will allow the new factors to be taken into account and if no steps are taken to communicate relevant experience. Had a fuller specification been given of the skills required on each shift—that at least two fillers and two pullers should be qualified to cut—a better distribution could have been achieved without interfering with customary methods. Such specifications were not given because it was part of the 'culture' to leave such matters in the hands of the 'tradition'.

Even a better initial distribution of experience and skills would have been of little avail if teams were still subject to random break-up when their faces finished. Cavilling rules which permit this are dysfunctional in composite longwall working.

A reorganization, less at the mercy of chance so far as team building and allocation were concerned, would have led to a more balanced and complete development of the composite system, with results substantially better than those obtained, however much these represented an improvement. The very wet face apart, five cycles per week would have been an appropriate aim for all teams, anything less being regarded as evidence of a degree of dysfunction requiring investigation and correction.

SECTION EIGHT

Preparing for Higher Mechanization

CHAPTERS

CHAPTER XXVII

The Need for a Machine-Centred Work Culture[1]

Throughout history the principal coal-producing agent has been the miner. Though various mechanical aids have more recently helped him to do his job, mining systems have still on the whole been designed primarily to make efficient use of human ability. Costs, productivity, and output targets have been largely determined by this need, as have face layouts and technical arrangements. The size of working groups and the customs and routines regulating mining practice have had the limitations of human capacity as an ever-present constraint. These elements together constitute a *work culture*. Despite the drive to increase mechanization, coal mining in Great Britain during the decade of the '50s remained under the domination of a man-centred or *manual work culture*.[2] With the machines that now exist, and still more with those under development, a rapid shift is taking place from the human being to the machine. In these systems no longer is the producer a man serviced by machines but a machine serviced by men. For the possibilities of this new situation to be realized a change is required in the work culture from a man-centred to a machine-centred attitude—a *machine culture*.[2]

A self-consistent machine culture will affect all dimensions of the work process, technical, economic, and socio-psychological. The implications for organization and manning were outlined in Chapter XI in relation to a concept of continuous mining. Certain additional aspects, of particular relevance to the present chapter, are discussed below.

[1] Source papers: 6, 19.

[2] For convenience 'manual' is used for 'man-centred' and 'machine' for 'machine-centred'.

THE NEW VALUE OF TIME

As coal-getting machines have a potential production rate of an entirely different order from manual methods (however much machine assisted), hold-ups, even when short, can mean substantial losses. A fifteen-minute stoppage on a machine producing a ton per minute loses 15 tons—a man-shift for a filler (cf. Chapter XIV). In a pit with 10 machines of this capacity working three coal-winning shifts in the 24 hours but with no arrangements to continue machine operations during bait stands, 4,500 tons would be lost per week. Time taken for men to travel from pit top to coal face can incur even more serious losses when coal-winning technique has attained this power. In the pits visited, effective shift time was rarely more than six hours and sometimes less than five. Haulage delays of 15 to 30 minutes, acceptable under hand-filling, become unacceptable under full mechanization. Recognition of the new value of time will fundamentally affect attitudes to all interruptions of the productive process—at any stage and through any cause.

MACHINE UTILIZATION

A corollary is a *new basis for productivity*. At lower levels of mechanization the natural basis is the man-shift—the time unit for measuring the work of the miner. With the machine taking over, the appropriate measure becomes the proportion of actual to possible machine time. At present, labour costs are still the major factor in production costs—60 per cent in 1958. As they diminish, an index of comparative productivity would be more appropriately based on machine utilization. How far optimum use is being made of machines is becoming a more fundamental question than how many men are on the face. In continuous process industries it is often found more economical to keep one plant operating 24 hours a day seven days a week than to have two or three plants operating only one or two shifts five days a week. In the culture of industries such as steel and chemicals the reality basis of continuous operation has been completely accepted and there is full commitment by all concerned to the consequential discipline in work behaviour. The plant is only stopped for planned maintenance, periodic major overhaul, or because of faults or breakdowns—or because bottlenecks still develop in certain parts of the process where the technology is not yet up to the level attained elsewhere. In a machine culture such an approach is axiomatic—and has indeed already been applied to mining for many years in a number of collieries, for example, in the United States, which have succeeded in remaining economic under severely competitive conditions. Though a great many pits will never be suitable for completely continuous operation, the degree of approximation to such an end-state can in the majority of cases be made much closer than it is at the present time, with higher machine utilization secured for whatever number of shifts per day and days per week may represent the optimum programme when all factors in the local situation are taken into account.

INTENSIVE MINING

A relatively small number of comprehensively mechanized faces can deliver a very substantial daily production, especially when worked continuously and concentrated spatially. One main road only is required and productivity is sufficiently high to absorb the capital cost of installing such facilities as a trunk conveyor, underground coal storage, an automated shaft-bottom, and skip-winding—the type of adjunct necessary if heavy and continuous production is to reach the surface without interference. Arrangements of this kind describe the configuration of *intensive mining*. The Reid Report (1945, p. 5) states that when face conveyors spread at the end of World War I, a concept of intensive mining began to emerge but failed to develop at a crucial stage in the history of the industry—especially in Great Britain. As a result, face districts developed, extensively, in all directions of the pit, while haulages continued to be very much as they had been, this framework being marginally sufficient for longwall working so long as hand-filling was retained. The persistence of such a framework into the post-war period has erected a major barrier against the full development of face mechanization, whose benefits cannot be realized while it still remains. The costs of change are enormous and can be met only from the increased productivity of the technology not yet fully in being. The initial failure, however, would not have occurred had the influence of a machine culture been stronger in the earlier period, and only its firm establishment at the present time can lead to the development and spread of intensive mining at the rate and on the scale required.

PROJECT ORGANIZATION

With the increasing variety of machines and equipment now available each face (or group of faces) has to be planned, set up, modified, and maintained throughout the span of its operating life as a specific project. Each such project requires a specification of its 'mission configuration' (Lyman, 1961) in all dimensions—technological, economic, and socio-psychological. This is necessary to an incomparably greater extent than when there was simply one prevailing method of mining, so wide has become the range of possible outcomes according to the way a given set of opportunities and conditions is handled. While project organization is most developed in recently emergent technologies such as those concerned with weapon systems, the basic assumptions are system characteristics of any machine culture. The concept is peculiarly appropriate to mining as precisely the same set of conditions never recurs. The increased speed of advance (or retreat) possible under full mechanization shortens 'face life' to a degree which makes the project character of underground work immediately perceptible. This situation can lead to great insecurity in the faceworker unless he knows well ahead what his next project is going to be and whom he is going to be with. Ideally, whole teams should transfer from one project to another with full preliminary discussion of the targets, nature, organization, and conditions of the new mission. Where a facework group can choose its next mission this is obviously beneficial. Where this is not practicable, it can nevertheless make some contribution of its own to the definition of the mission configuration and to its re-appraisal as work proceeds. Without this, it is doubtful if full commitment can be secured and such commitment is essential in a fully developed machine culture.

CHAPTER XXVIII

Mechanization in a Manual Context[1]

The limited extent to which a machine culture had yet taken hold may be inferred from an N.C.B. Information Bulletin published towards the end of the research period (No. 58/194, 1958), which gave the mean o.m.s. of mechanized faces as 5·6 tons, against a national average of 3·3. This is within the uppe limits of the partially mechanized range and may be taken as a gross indication of the extent to which new machines were being used in a manual context.

When a mechanized system is introduced in a pit or seam where the dominant work methods belong to the traditional manual culture, one may expect it to be operated in terms of the norms and values of this culture rather than those of the more appropriate machine culture. When mechanized systems become dominant, as in some of the pits visited in East Midlands, the conditions exist for an acceptance of the expectations and levels of productivity of a machine culture. As yet the new frame of reference had little influence on the pits visited in North West Durham.

A HAARMAN SCRAPER-PEELER IN A MANUAL CONTEXT

The treatment of this machine—the first to be introduced in Great Britain—is typical of what happens under these circumstances, though the project was highly successful from other points of view (*vide* p. 270). Installed in a seam containing several conventional cutting and hewing faces, it was referred to as 'the cutter' and its daily advance was set at 4′ 6″—the length of the cutting jib customary in the pit. Capable of producing over 30

[1] Source papers: 6, 19.

tons an hour under the conditions obtaining, it ran for no more than three hours in the 24 with the result that its production was of the same order as that of the conventional longwalls—though a little better. Overall cycle time was reduced to two shifts but the shift gained in this way was left as dead-time in order to fit in with winding arrangements.

Face layout limited machine advance to the gateway advance achieved by hand on a separate shift. Yet simple and proven alternative methods would have enabled stonework and facework to be done simultaneously and would have made a second coal-getting shift possible (cf. the Dutch arrangements described in Chapter X). Later, the introduction of scraper-packers, though permitting a somewhat greater and more rapid stonework advance (5′ 0″), did not result in a longer time for the machine run. The new equipment was viewed as affecting stonework only, not the face system as a whole.

This low level of machine utilization is typical of what happens when machines are used in a manual culture. In a machine culture, allowing high-producing machines to remain idle while manual operations catch up would be experienced as intolerable. On the Haarman face gains were reckoned in terms of savings in manpower; and, indeed, throughout the research period reports in the mining press—which reflected the general outlook of the industry—quoted face labour costs as the most usual indicator: if these could be reduced and manpower saved, even if total output was not increased, the system was regarded as acceptable. This attitude prevailed in the pits visited.

A MODEL FOR USING A HAARMAN SCRAPER-PEELER IN A MACHINE CONTEXT

The kinds of steps required in order to carry out facework in terms of a machine culture, with consequent maximization of machine time, are exemplified in *Table 33*.

In *Table 34* these principles are applied to the Haarman installations, and *Figure 19* contrasts this model with that operating in a manual context. Even with the not unusual restriction of two-shift winding, three machine periods are none the less achievable. Assuming 30 tons per machine hour and seven

TABLE 33 METHODS FOR INCREASING MACHINE TIME

Step	*Effect*
1. (a) Carry out stonework simultaneously with facework.	(a) The gain of an additional coal-winning shift.
(b) Where conditions do not allow, add additional men to caunch teams, especially in the mother-gate, and take advantage of any partial overlap possible with work in the face.	(b) The reduction of time spent on a separate stone shift.
2. (a) Carry out advancing simultaneously with machining.	(a) The gain of an additional coal-winning shift.
(b) Where conditions do not allow, add additional teams for chock-drawing, taking advantage of any partial overlap possible with machining.	(b) The reduction of turn-round delay with buttock getters and of the advancing phase of a coal-winning shift with frontal getters.
3. Add a fourth shift in the 24 hours by staggering times so that the next team arrives at the coal face immediately the previous team has finished.	(a) The recovery of time lost through travelling from pit-top to face. (b) The elimination of dead periods during the 24 hours.

interchangeable composite teams of 10, the expected o.m.s. would be in the neighbourhood of 13·5 tons, whereas that being attained by the continuation of conventional arrangements was in the neighbourhood of 4·0 tons.

TABLE 34 ACTIVITY SEQUENCE FOR INTERCHANGEABLE COMPOSITE TEAMS

Shift Time	*Team*	*Shift Period* — *Phase One* *Activity*	*Hrs*	*Phase Two* *Activity*	*Hrs*
0.00 hrs	(a)	Machine	4	Advance	2
,, ,,	(b)	Stonework	3	Stonework	3
4.00 ,,	(c)	Advance	2	Machine	4
6.00 ,,	(d)	Stonework	3	Stonework	3
10.00 ,,	(e)	Advance	2	Machine	4
,, ,,	(f)	Advance	2	Stonework	4
16.00 ,,	(g)	Advance	4	Stonework	2

For convenience shift times are given from 0.00 hrs. Though the underground shift is 7½ hrs, with 30 mins each for travelling in and out and another 30 mins for bait stands (which separate shift phases), effective face time is 6 hrs. The overground shift is 8 hrs, which gives 16 hrs winding time, extendable to allow some margin.

FIGURE 19

MODELS FOR A HAARMAN INSTALLATION

above, in a machine context

below, in a manual context

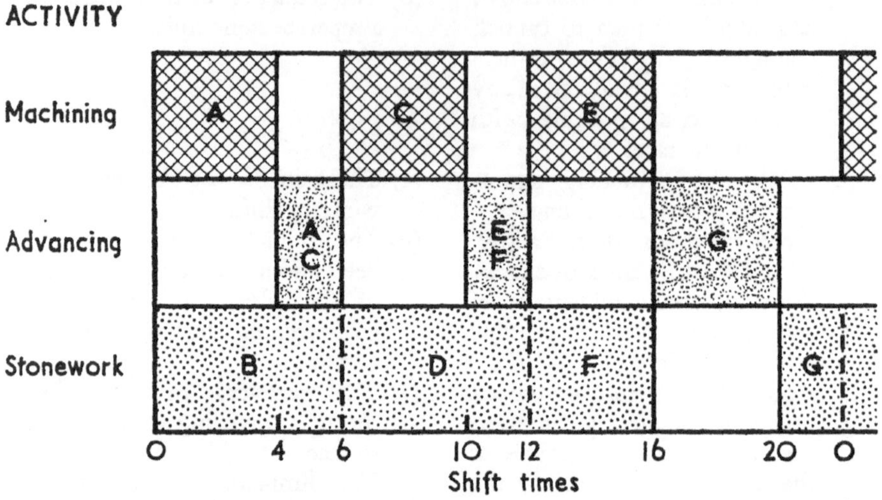

(A–G : 7 interchangeable composite teams of 10 men each)

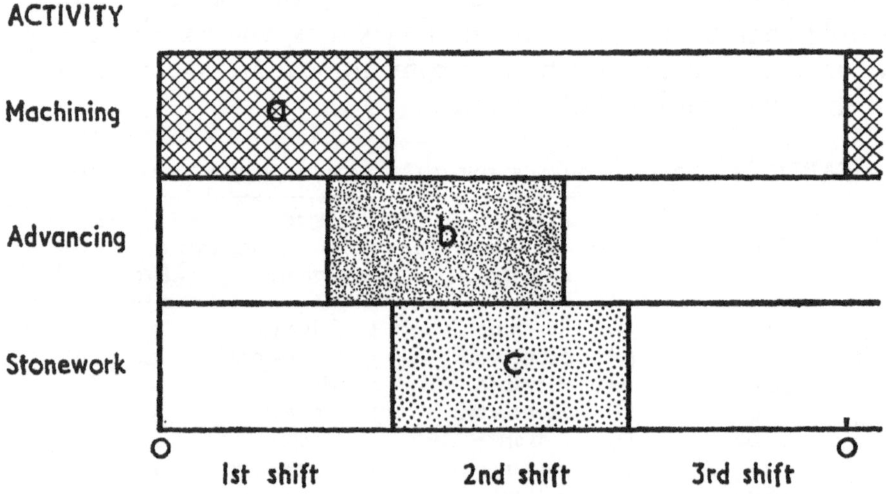

This estimate is based on an effective shift of 6 hours, divided into a machine phase of 4 hours—which contains an allowance for interference of half an hour—and an advancing phase of 2 hours in which double teams are employed. Shift times are so arranged that team (c) works coincidentally with team (a) during its advancing phase. After a joint advancing phase which begins the third shift, team (f) goes on to stonework, while team (e) takes over the final machine run. A single late nightshift completes chock-drawing and gateway packing in order to make the face ready for machining the next day.

The model has assumed, therefore, that conditions would allow concentration on advance by two simultaneously operating teams (a) and (c), and (e) and (f)—beginning at 4.00 hrs and 10.00 hrs respectively—with five pairs taking chocks on each side (there was no belt). With no nightshift winding, one advancing stonework team would suffice on the last shift.

The model has also assumed that conditions would allow stonework to proceed simultaneously with facework. It would be necessary for the mothergate caunches to keep up, but the tailgate men would always be available as temporary reinforcements to either the machine or advance teams. Within reasonable limits, lag in the tailgates could be eliminated as occasion arose.

A further assumption is a unit team strength of 10, based on existing arrangements for machining. This may be more than would be required on stonework with slushers, but some reinforcement would be necessary if the chocks became difficult and threatened the loss of machine time—more important to avoid than to achieve a small saving in manpower.

When the research team was present, machine time averaged 3 hrs for a 5′ 0″ advance. An improved haulage, however, was expected to increase machine time by reducing external interference. There was also some possibility of lengthening the face, deepening the advance, and raising machine speed. The estimated new machine time of $3\frac{1}{2}$ hrs (with an extra 30 mins allowed for interference, to make up a shift phase of 4 hrs) is conservative, and a figure of 13·5 tons is by no means the highest estimate of possible o.m.s. that could be made. For example, it was already the practice to begin drawing off at the two ends of the face before machining was finished. This overlapping would be a crucial feature to develop if a longer machine run became possible and the roof allowed a 6′ 0″ advance per machine period. The objective would be to reduce the interval required for advance to as short a time as conditions would permit. Moreover, if one advance team started 30 mins ahead of the other, bait times could be staggered and work could go on without interruption. With such developments it might be possible to gain 30 mins extra machining at the end of the shift. This would add 45 tons to the day's production—more on a lengthened face—for no additional labour.

A model of this kind calls attention to the discrepancy between potentiality and attainment, and shows that the reasons lie in the persistence of procedures associated with a manual outlook. Whatever difficulties might be encountered in implementing the scheme suggested, obtained results should be nearer the new standard than the old. The attainability of a level of productivity outside the range of existing norms would be demonstrated.

RESISTING CHANGE

The transition from a manual to a machine culture will take place only if the resistances aroused by the need to accept a radically different frame of reference are understood and handled constructively. There were many indications in the pits visited of unwitting use of techniques of psychological defence to circumscribe and damp down the consequences of change. To gain insight into the nature of the resistances represented by these defences was exceedingly difficult for those involved, in view of the unanimous character of the local rationalizations, which pervaded the entire mining outlook. The resistances were stronger because of the age of the coalfield with its established culture, the complicated emotions aroused by the experience of the working-out of pits, and the widespread doubt that effective plans for alternative employment would be made—in time, or at all.[1] In one colliery particularly, a common subject of discussion was how to make the coal last, rather than how to increase production: schemes for technical improvement gained general acceptance only to the extent that they warded off the more immediate

[1] More recently NCB has taken far-reaching steps on transfer and resettlement.

threat of the pit becoming uneconomic and being closed down.

A defence encountered on several occasions was that of restricting the trial of new equipment to conditions where standard practice could not be followed. These conditions would often 'happen' also to prove unsuitable for the particular equipment tried, though it would be said later that it might have done better than the standard in other conditions. Another technique was to keep change projects down to such a small scale that their effects got lost in the larger activities of the pit. Legitimate concern with safety could also be used as an unconscious psychological defence, as when a hazard which occurred during a phase of retreat mining in a rather gassy pit led to a directive which was taken as a reason for not trying it in a safe pit.

Where successes occurred with new methods they would often be discounted as special cases and so much attributed to the idiosyncrasies of local circumstances that no attempt would be made to conceptualize the experience. As no 'general lessons' had been learnt the question of further applications scarcely arose.

By far the most common defence, however, the more difficult to recognize because it contained a positive element, was to proceed with a mechanization or organizational project—and get rather better results than with existing practice, yet not too much better. While reasonable results would show that progress was being made, too good results would demonstrate that a different and altogether more highly productive order was an attainable reality. This would threaten the basis of existing norms, which there was need to preserve as a sign that the established adaptation of the enterprise—and the working community dependent on it—to its environment was still viable. This is one meaning of the productivity figures quoted from the N.C.B. Information Bulletin: they show a respectable improvement without making the case for radical change. The realism of this attitude is that it allows the existing framework to be modified with a minimum of disturbance to the internal system of the industry. Its unreality is that by slowing down the rate of change it increases the threat from uncontrollable forces in the external environment, such as a competitive oil industry, so that in the end there is no gain in security.

CHAPTER XXIX

The Selection and Training of Face Teams[1]

AN EXPERIMENT IN TEAM PLANNING

It was suggested in Chapter XI that existing customs in North West Durham for the self-selection of work teams and their allocation to faces by cavilling were not consistent with the most effective working of more highly mechanized systems. This was recognized by both management and lodge when Haarman scraper-peelers were introduced. Both sides were anxious that the new units should succeed and a frank meeting produced a decision to replace the traditional marrowing up of teams by a new and more appropriate institution. For the first units the lodge agreed that, from among those volunteering, management would have the right to include certain men rated 'excellent' on mutually acceptable criteria and to exclude certain others rated 'poor'. On its side, management agreed that the middle and largest group, rated 'good', should go into the cavil; from these the balance would be made up. In this way the quality of the teams pioneering the application of a technology new to the British industry was guaranteed.

Before the introduction of the Haarmans there had been seven hewing groups in the seam, notoriously different in their productivity, no matter to which face they were cavilled. The best gave an o.m.s. (hewers only) of 5–7 tons, the worst no more than 2–3 tons. With the poorer men excluded from the Haarman faces, hewing performance began to deteriorate and management and lodge came to a further understanding over a proposal to introduce cutting with flight-loading into the remaining faces. A selection committee of four was set up, consisting of the pit manager and the seam undermanager together with the president

[1] Source papers: 5, 19, 23.

and secretary of the lodge. This committee recommended not only which men should go to which faces but saw that key roles on all were manned by competent and reliable workmen. The committee reviewed the performance records of all faceworkers, taking into account the range, level, and appropriateness of skills, reliability in attendance, willingness to co-operate on the job, capability under difficulties, etc. They built up teams accordingly. Though final responsibility rested with the manager, the lodge had a full say in deciding both the principles on which teams were made up and how individuals should be allocated. Optimum use was made of the skill resources available, a balance of skill between and within teams was assured, and no 'problem' team remained. The results were accepted by the men. This whole development—which is in striking contrast to the seam re-organization described in Chapter XXV—illustrates the need for planned teams in the organization and manning of mechanized faces which was discussed in Chapter XI.

MAKING USE OF EXISTING STRUCTURE

In another pit where flight-loading was introduced on composite longwalls the technological innovation was not so radical and existing skills could meet the new demands. The problem was to select from a well-knit and highly successful 41-man team (the No. 2 Panel described in Chapters XIII and XIV) a group of 25 to go on to the first flight-loading face. The team was invited to marrow itself into a set of 25 men, which meant excluding 16 of its members and dividing into a 'have' and a 'have not' group. The suggestion of one of the team captains that the men should cavil for places was not followed and various men made independent 'bids' to form the necessary set. Relations, hitherto harmonious, became severely strained and those who had not succeeded in joining the set began to refer to the others as 'the very special supermen'.

For reasons yet to be discussed (p. 274), the plan was not implemented. Having, however, rejected 16 out of its 41 members, the team took many months, while remaining on hand-filling, to overcome the ill-feeling generated. A year later management was able to begin its first operational flight-loading longwall. Two separate flight-loading faces, each with 22 men, were planned, which meant that all team members who wished could transfer. The team divided along the natural boundary of its shift-rotating groups. These, after consultation with management, added one or two men to make up the required numbers and then cavilled to determine which face each should work.

The two flight-loading faces were laid out as a double unit and both teams were paid on the same national Power-Loading Agreement. They insisted, nevertheless, on working as single units, since they believed that, should flight-loading fail and the faces revert to hand-filling, this would give them a better wage. This was because the latest single unit prices in another seam had been somewhat better than existing double unit prices (cf. Chapter XXI). For optimum working of a double unit close collaboration between the two faces is necessary; a common note helps to ensure this, separate notes being more likely to inhibit flexible deployment advantageous to both sides, especially when different conditions develop on the two faces. In the event, the teams helped each other in much the same way as if they were still a double unit, the hewing and stonework in the mothergate, for example, being jointly undertaken. It is unlikely that this effective relationship would have developed had the 41-man team been obliged to reorganize along lines inconsistent with the structure it had evolved—as the tension created when this was threatened forcibly suggests.

The emotion released by the well-intentioned early plan indicates the importance, when introducing higher mechanization, of respecting and utilizing not only the outward tradition but the deeper-lying social structure and origins of existing work groups. Each new situation is likely to have its own best solution, which must take into account not only factors within face teams, but, as the Haarman units show, considerations affecting the seam, and indeed the pit as a whole. Otherwise the transition from a manual to a machine culture will fail to take place.

TRAINING IN A NEW TECHNOLOGY

The introduction of new kinds of equipment creates learning problems. With the simpler kinds, such as scraper-packers, deputies and overmen acquire the necessary know-how from their own manager, from visits to other pits, or at Area schools. They then instruct the men. More complicated equipment, particularly when under operational trial, is installed by an Area development team, which tries out the machine with attached trainees who then gradually take their place. One Area man remains for a time to act as supervisor.

Other tasks belonging to the cycle are carried out by local men who continue to work with those newly trained when the production phase is reached. The Area team is concerned exclusively with its own specialist task, having no responsibility for the rest of the cycle, which remains with the deputies, overmen, and seam undermanager. The isolation of the machine activity in this way fits the pattern of expectations associated with the conventional longwall, making the development team into a segregated task group which requires external control. The question arises as to how far a method of training based on such a group will perpetuate inappropriate practices.

FLIGHT-LOADING (CASE ONE)

Two studies were made of the introduction of flight-loading in the pit in which composite longwall working had emerged. The first was an experimental single unit designed to test the feasibility of flight-loading in the Manley. The face was cut by a machine fitted with a turret jib—which eliminated the need for shotfiring—and an automatic gummer—which made scuffling unnecessary. For filling, the cutter was fitted with Lambton 'flights'. The machine was operated during the cutting and loading shifts by an Area development team of four. To this nucleus were added spare men and back-bye workers who carried out the pulling and stonework. This *ad hoc* arrangement came into being because management could not risk removing an established team from one of the on-going and highly productive composite panels. The *ad hoc* groups worked as specialists, for it was the mining method which was on trial. As the Area team, however, trained only one task group to operate the equipment, only this group was in a position to practise task continuity; the pullers and stonemen could not reciprocate, neither could they take part in task-shift rotation. The training phase, therefore, produced a pattern which could create fission within a composite team coming on to the face.

After six weeks the face was stopped because of a limited market for the type of coal worked, much to the dismay of all concerned. It was not started again until a year later when it was manned by a different team—the experienced composite group described in Chapter XIII. Yet management planned that they should work as specialists. There was to be no rotation of tasks, though quarterly changes might be possible in the future. This was the pattern which had been used on the experimental unit. The team, supported by the lodge, did not want to break up the composite system they had developed. The manager, however, would have no rotation between cutting-loading and pulling-stonework for at least two months; thereafter, he thought men on the cutters and flight-loaders could move off on to pulling-stonework two at a time at fortnightly or monthly intervals.

Nevertheless, from the early weeks the teams operated a task-shift rotation procedure, based on their previous experience, without causing any change in the exceptionally high level of productivity. The undermanager was aware of these developments but judged the teams capable of managing themselves. The manager had underestimated the team's intuitive capacity to adapt its earlier composite experience to the new technology and, being concerned for the success of the enterprise, had wanted to use the conventional model which had been created round the Area development team.

FLIGHT-LOADING (CASE TWO)

In a different part of the same pit an attempt to introduce a multi-jib cutter-loader had been unsuccessful because variations in seam height made it difficult for the machine to traverse the face. Flight-loading and scraper-packing were then tried under the aegis of an Area team and the face was taken over, under supervision, by a volunteer group of composite workers from several production and development faces. The same kind of difficulties were encountered with the flights as with the multi-jib cutter-loader. Moreover, the stonemen took a long time to settle down with the scraper-packer. The lodge considered the team too small; it also thought it would have been better to have introduced the flight-loading first, allowing the team to become familiar with it before introducing the scraper-packer.

During the experimental phase team members were allocated to permanent roles and then, since the cycle was often out of phase, assigned according to need. This pattern, which became established, was followed when the face became operational. The team protested that they were being too closely supervised by deputies and 'outside' officials and were not being given a chance to apply their composite experience. They resented the introduction of reinforcements and special men to do jobs they could have done themselves. Even re-deployment during a shift was carried out by the deputies. The persistence of this pattern of external control induced a crisis which was only resolved when conditions made the continuation of flights impossible and the machine was withdrawn. The team was then allowed to organize itself, and the face, with hand-filling and scraper-packers, rapidly settled down to completing five cycles a week.

These cases suggest that on-the-job training restricted to the technical preparation of one group for a specific operation tends to crystallize work organization around the training team and to inhibit the development of forms more appropriate to the cycle of operations as a whole. Where the immediate face management accepts the capacity of the team to adapt and apply its relevant experience there is less danger of premature crystallization of inappropriate work methods.

AN AREA SCHOOL

In view of their close contact with local work groups, Area development teams have unusual potentiality as change agents. To be successful, however, they must consider any particular machine in relation to the cycle as a whole—and in socio-technical rather than purely technical terms. Such issues, however, are the concern not only of the Area group and the men assigned to the experimental face, but of local management, from the deputies upwards. The introduction of higher mechanization involves a 'vertical slice' of the pit, and colliery-Area relations as well as

TABLE 35 COMPOSITE WORKING WITH THE INTRODUCTION OF A CUTTER-LOADER

A FIRST COURSE FOR DEPUTIES, TEAM CAPTAINS, AND MACHINE OPERATORS

(Each course made up of a party of eight from each pit in turn, comprising 2 deputies, 3 team captains, and 3 machine operators)

Time	*Event*	*Official or Group Responsible*
Monday		
9.0 a.m.	Arrival of course group, settling in, and issue of specially prepared notebooks.	Course Tutor
9.30 a.m.	Introductory address touching on the need for technological development, the history of the present project, and future Area policy on mechanization.	Area General Manager or his representative
10.0 a.m.	Lecture-discussion on the system as a whole—concentrated workings, the new face, hand-filled and machine-operated versions—with models, sketch plans, etc.	Course Tutor
1.30 p.m.	Lecture-demonstration on the surface of adjustable props, new timbering system, and flexible conveyors.	Course Tutor
3.30 p.m.	Lecture-discussion on the social organization of new faces.	Lodge Secretary, demonstration pit
Tuesday		
9.0 a.m.	Course group received and briefed at demonstration pit.	Manager
10.0 a.m.	Taken to hand-filled faces in groups of four.	Face teams
12.30 p.m.	Group discussion on the surface.	Manager, Deputies, Lodge Secretary, Team Captains
Wednesday		
9.0 a.m.	Evaluation of pit visit and hand-filled system.	Course Tutor
10.0 a.m.	Lecture-discussion on machine-filled face, stressing the organization of the face team, employing scale models.	Course Tutor
1.30 p.m.	Demonstration-discussion on the surface of full-size model of cutter-loader.	Course Tutor
3.30 p.m.	Visit to Area Workshops to inspect and discuss actual machines.	Area Mechanization Officer or his representative

Time	*Event*	*Official of Group Responsible*
Thursday		
9.0 a.m.	Course group received and briefed at demonstration pit.	Manager
10.0 a.m.	Taken to machine-filled faces in groups of four.	Face teams
2.0 p.m.	Group discussion on surface.	Manager, Deputies, Lodge Secretary, Team Captains
Friday		
9.0 a.m.	Evaluation of pit visit and machine-filled system.	Course Tutor
10.0 a.m. – 5.0 p.m.	Descend demonstration pit for full working shift, taking part in facework. Group discussion in period between fore- and backshifts.	Foreshift Deputy, Team Captain, Machine Operator
Saturday		
9.0 a.m.	Discussion group on experience of developing and operating the new system.	Area Production Manager or his representative
10.30 a.m.	Course group joins regular Saturday morning review meeting of representatives of pit groups concerned with the new system.	Course Tutor

those between the working group and local management. The size of the implicated system has become very much extended. Lack of a common approach and mutual understanding creates difficulties and delays and increases tensions and resistances.

To help overcome such problems the research team assisted in the development of an Area school, at which faceworkers and officials engaged in new mechanization projects could be given a total picture of the programme and an opportunity not only to learn new skills but to clarify emergent roles and relationships. Courses at the school were informal and conducted in small groups. They included systematic exposition of new principles from the point of view of organization and manning as well as of mining; familiarization with new machines and equipment overground; working periods at demonstration faces; group discussions with those concerned at all levels; and course evaluation periods. Managers, undermanagers, and overmen first went through a course in their own groups so that problems at each level could be identified and discussed. The colliery managers then held conferences with those of their officials who had been through the course to make a local plan and secure lodge agreement. They next selected deputies, team captains, and faceworkers to attend a week's course at the Area school in groups of eight. These teams became internal development groups in each pit, going back for extra practical periods to the demonstration faces and joining in weekly discussions arranged by Area on problems of the new systems. As soon as the groups had made a successful local adaptation, they began training other teams who, having attended a course on 'principles' in their own pit, would work with the local development group before starting their own faces. Area personnel were available on request for attachment to local groups or on the initiative of Area if, after review, this appeared necessary.

Table 35 gives a specimen syllabus for a local development group. The course presented composite working as a means of moving towards continuous mining; a cutter-loader was then introduced into the new frame of reference. Men already familiar with the hand-filled model spent more time at mechanized faces. Local groups met those responsible for every aspect of the new system and had ample opportunity to express misgivings and to ask questions.

As always in the experience of the research team, men and officials alike showed unusual capacity to make constructive use of group methods of learning as soon as an opportunity was afforded. According to the reports of the trainees and the impressions of the staff, the course produced a massive shift in attitudes—away from a manual and towards a machine culture. Unfortunately, the research team was unable to pursue this study to the point of developing measures for the systematic follow-up of a series of the groups who passed through the school.

CHAPTER XXX

The Development of Adaptive Work Organization[1]

THE PROGRESSIVE AMALGAMATION OF TASK GROUPS

The appositeness of composite organization for higher mechanization, confirmed by the subsequent record of the flight-loading faces just described, receives further support from the spontaneous efforts made by management in another pit to proceed by stages towards a composite form of organization. Although production longwalls were conventionally organized, two development faces—one hewing and one cutting—had been successfully worked as composite bargains. A third development face had been prepared to open up a new area of the pit where production faces were ultimately to have flight-loading, power-pulling, and scraper-packing. Flight-loading was not used on the development face, since this would have meant giving considerable time to technical experimentation which would have slowed up the rate of development.

The development face was 55 yds long and was undercut to a depth of 5′ 0″ with a machine fitted with an automatic gummer. The shortness of the face allowed the activities concerned with coal preparation to be carried out by one group of two men, while a team of five undertook the filling. Because of the largeness of the gateways two scraper-packers were installed and the face was completely stowed; the slusher winch was also available for power-pulling the face conveyor. A single team of eight 'gateworkers' sharing a common note became responsible for all activities previously carried out by pullers and stonemen. In

[1] Source paper: 4.

manning this face the manager was able, under a new power-loading and packing agreement, to select men for cutting and scraper-packing. The filling team—because it was not using power-loading—continued to be allocated each quarter according to cavilling rules.

Both on the development face and on the 250-yd double unit production faces which came into operation later, management and lodge advocated that the cutting group should combine with the fillers and share their earnings equally as a composite group, paralleling that formed by the pullers and stonemen. The manager offered to pay such a team a single tonnage price. The proposal proved too radical for men whose entire experience had been in conventional working, and cutters and fillers remained on their own notes.

With further technological change, a fusion took place under the power-loading agreement and there were now two groups, one concerned with preparation and flight-loading and the other with power-pulling and scraper-packing (and also chock-drawing where the goaf was not stowed). The fusion of these two groups into an overall team was held up by their lack of training in each other's equipment; but the advantages to be gained from task continuity would have more than compensated for any required training time. These experiences, though only partially successful, suggest that successive fusion of conventional task groups is a model which those introducing higher mechanization might use as a point of departure (Appendix III).

CHANGING TO COMPOSITE ORGANIZATION IN THE DONBASS

At the time the research was carried out not many sets of data were available which gave quantitative information on the effects on productivity of converting mechanized faces from conventional to composite organization. One of the best designed experiments—in which the same men continued as a composite team on a panel which they had previously worked as separate task groups—was carried out on the 24th Western face in a colliery in the Donbass. The results, together with those obtained by the team when they later moved to another face, are set out in *Table 36*.[1]

TABLE 36 RESULTS OF CHANGING FROM CONVENTIONAL TO COMPOSITE ORGANIZATION BY A DONBASS FACE TEAM

	When operating as Separate Task Groups		*After re-forming as a composite team*				
	24th Western Face				*25th Western Face*		
	Oct. 1952	Nov. 1952	Dec. 1952	Jan. 1953	Oct. 1954	Nov. 1954	Dec. 1954
Task Dimensions							
(a) Length of face (yds)	132	134	140	142	141	134	128
(b) Seam thickness (ins)	47	48	46·5	46·5	41	41·5	41·5
(c) Depth of cut (ins)	55	57	57	57	57	57	57
Daily Output							
(d) Target (tons)	310	308	310	320	295	320	334
(e) Actual (tons)	309	338	360	337	387	342	347
Cycles Completed							
(f) Standard	31	28·8	30	30	31	33·6	37·2
(g) Actual	31	28·8	32·9	30·3	39	37·1	39·2
No. of Men at the Face							
(h) Planned	48	49	48	45	41	41	42
(j) Actual	45	49	41	41	38	38	43
(a, b, c) Task Size (cu. yds per cycle)	263	283	286	291	254	245	234
(g/f.h/j) Completeness/ Manpower Index %	107	100	128	111	136	119	103
(e/j) Face o.m.s.	6·85	6·93	8·67	8·12	10·2	9·0	8·1

The average face o.m.s. on the 24th Western face prior to the formation of a composite team was 6·9 tons; afterwards it rose to 8·4 tons, an increase of 22 per cent, as against an increase of almost 6 per cent which could have been expected from the increase in the task size. When, 18 months later, the team went on to open up the 25th Western face, the average for the first quarter was 9·1 tons, an increase of 32 per cent, although the task size decreased by 11 per cent. On two faces in another pit

[1] The Russian source, Ugletekhizhat, Moscow, 1956, was drawn to the attention of the research team by the Department of Scientific and Industrial Research, who made available translation facilities. The experiment is also summarized in the Report by the Technical Mission of the National Coal Board, Part 2, 1957. *Table 36* re-arranges the original data and adds calculations of task size and the completeness/manpower index.

where composite working was first tried, increases of 10 per cent and 14 per cent were obtained—in the latter case despite a loss in coal height of 4·4 per cent and in face length of 2–3 per cent. In all reported cases increases occurred.

The data given in *Table 36* on planned and actual cycles completed and face manpower permit the calculation of a completeness/manpower index broadly comparable to that used in Chapter XXVI. On the 24th Western face the mean index moves from 103 per cent prior to composite organization to 119 per cent after, an improvement of the same order as found in the Bramwell study. The downward trend of the index over three months on the 25th Western face does not, however, signify a decline in performance: rather it highlights the fact that after two years' experience with composite organization the actual level of performance (in terms of cycles completed) had come to be accepted and implemented as a new standard or norm to be expected. Against the starting standard of 31 cycles, conceived as an improved norm within the conventional culture, the index for these three months would have averaged 132 per cent. Indeed one might observe that when such a performance index remains considerably in excess of 100 over a period of time, this may be taken as an indication that not only do the targets need revision, but that there has been no real learning and acceptance of the experience of working within a new culture requiring its own norms. Such appears not to have been the case in the Donbass.

Yet the composite organization of these teams remained seriously incomplete (a rating on the scale used in Chapter XXV would be no higher than 8). Still better results would undoubtedly have been obtained had not those concerned with caving, gateway stonework, and mechanical repairs remained outside—'on call'. Much time was wasted over this while the face team itself went on to shift work.

These developments in the Donbass appear to have taken place at about the same time as mechanized composite working appeared in East Midlands Division on faces visited by the research team. Here, composite organization was complete, but as these faces had been worked on composite lines before cutter-loaders were introduced similar comparisons could not be made. The

Russian data—from an independent source and a different cultural setting—confirm the findings of the research team in North West Durham and are in line with the theoretical expectations arising from the main hypothesis put forward concerning the nature of composite work groups.

GENERAL CONSIDERATIONS

In several of the cases studied during the research social change appropriate to technical change was apparently achieved only to be finally resisted. Absent on each of these occasions were techniques of effective communication between management and lodge officials—who had reached understanding among themselves—and the men directly concerned. The ordinary lodge meeting, where attendance is chancy, cannot always be depended on as a means of resolving the difficulties likely to arise in change situations. Outcome may be only too easily determined by prejudiced sub-groups who have had no contact with a particular innovation or who imagine their special interests to be threatened.

Direct discussion is required between representatives of management and lodge and the men to be affected. If many men are involved several discussion groups are required, each of face-to-face size, as well as a general meeting. By such procedures, as has been shown in other contexts, feelings can be ventilated and explanations made—with a chance of full understanding (Lippitt, 1949; Thelen, 1954; Trist & Sofer, 1959). As a result of such informal discussions, decisions at lodge meetings are likely to be at a higher level of reality, as are proposals put forward by management. Unfortunately, no tradition was available in the mining community which would allow such discussions to take place as a matter of course. Yet whenever they were attempted, as with the men going on to the Haarman faces, they were successful, and those present—on both sides—rapidly learned the social skills relevant to such occasions.

The case studies suggest that change from a manual to a machine culture requires some kind of 'unfreezing' of the existing situation (cf. Lewin, 1951) before the new culture can build up. Change towards a more advanced technology, however limited, provides such an unfreezing event. Technical change cannot, however, be of great value for this purpose unless its potentiality for inducing social change is recognized. Those responsible for the detailed planning and initiation of technological change are not unnaturally concerned with the technical efficacy of the proposed systems, particularly in view of their high capital cost and the extent of the pit reorganization necessitated by more advanced schemes. Unless, however, a new system is capable of effective socio-psychological functioning, technical improvements may be of little avail.

This research has shown that in most technical situations there are possibilities for different kinds of work organization and that the alternatives vary in the extent to which they help or hinder the completion of the primary task. It is important, therefore, to identify the characteristics which facilitate this objective. Traditional single place organization, a product of generations of underground experience, contains all the main characteristics required of a social system to meet the demands of facework. A high level of positive orientation to the primary task is ensured by the fact that the men in each work group make up a single social and economic unit; they share one paynote and one job, which is the whole primary task. There have developed customs such as self-supervision, task continuity, and role rotation which are optimum attributes for work groups in a high risk situation. With the development of partially mechanized longwalls, this tradition was abandoned in favour of specialized work roles and segregated task groups. By its very nature, this type of organization cannot give rise to a social system in keeping with the demands of the coal face environment. Yet it has been carried forward into the type of work group recently developed for single task machines. Unless there is awareness that different types of group are possible, there is danger that the conventional approach will persist into the organization of even the most comprehensive mechanization projects. So far as this happens, the chance will be lost of correcting the harmful effects which have resulted from the use of an inappropriate 'mass production' model in the original change from single place to longwall working.

On some of the composite longwalls studied in this research principles of work organization deriving from the single place tradition were successfully applied in the new situation of the longwall coal face, with workable solutions to the many problems of reshaping them to an entirely different technology. The existence of this model makes easier the task of reinterpreting these principles once again to fit the requirements of a third situation—that which obtains as full mechanization is approached. The difficulty of maintaining specialized work roles and separate task groups in continuous mining and the general application of the Power-Loading Agreement provide conditions which favour thorough practical exploration of the adaptiveness of composite work groups for future facework organization. Under the impact of increasing mechanization, job enlargement and role de-differentiation are replacing job breakdown and role segregation in a number of industries. It remains for the coal industry to rediscover and reinterpret an invaluable tradition persisting from its own long and varied past.

Summary and Conclusions

Summary and Conclusions

This research programme has been concerned with the interaction of technological and socio-psychological factors in industrial production systems—represented by a variety of mining methods at differing levels of mechanization under low seam conditions. There were three main systems: traditional single place working; longwalls in which coal was cut or hewed and hand-filled on to conveyors; and longwalls at higher levels of mechanization. On the longwalls two radically different forms of work group organization were studied—the conventional and the composite. The research design involved the intensive case study, both qualitative and quantitative, over an extended period of time, of the structure and functioning of the social system associated with at least one example of each of the main mining methods, together with two or more subsidiary studies.

The pattern of work organization—the way in which those who carry out the necessary tasks are related to each other—was analysed in terms of the quality of work roles, the kinds of task groups, the prevailing work culture, the nature of the inter-group relations, and the character of the managing system. In order to compare the operational effectiveness of face systems of different types, two frames of reference were constructed. The first assesses methods of cycle regulation and defines concepts of cycle state, interference, and counteraction. The second—concerned with the outcome of the regulatory processes—assesses actual in relation to potential production performance and defines concepts of output level, task situation, and group response.

In single place systems, the miner possesses the necessary range of skills to undertake all facework tasks in a self-contained work place. His role is that of a multi-skilled, self-supervising workman towards whom the deputy stands in a service rather than a supervisory relation. Groups of up to six men share a place, the men selecting their own mates. Since all members do all jobs, either on the same or on different shifts, they share equally in the same paynote. Traditionally, a miner achieves faceworker status through being trained in a particular seam in which he holds a series of jobs on the haulage before reaching the face and in which he has a number of established rights. In general, relations between the men in the groups and between the groups themselves are harmonious. Such a pattern of work organization is well adapted to the technological conditions of single place working and to the general hazards of the underground environment.

Conventional longwall systems represent a sharp break with single place working, more so in the case of cutting than in hewing longwalls. In contrast to the one all-round work role, there is a formal division of labour with specialized tasks carried out by a number of groups of varying size. The larger task groups are identical role groups in which all members are supposed to do the same amount of the same task; relations in such groups tend to be troublesome when differences in capacity and willingness become apparent. The smaller task groups are reciprocal role groups in which each member makes a unique and interdependent contribution and in which relations are more amicable, especially as such groups are self-selected. There are also isolate roles which, apart from the shotfirer who is an official, are not those of a full specialist. Each task group has its own customs and agreements, including separate paynotes, so that each is segregated from the other and bound within its own field of interest. Since the groups do not spontaneously work together, co-ordination and control have to be provided entirely from outside—by management. Traditionally, however, the role of deputy is one of service rather than of operational leadership; coercive control, on the other hand, would be unsuitable and impracticable in the high risk situation of the coal face environment. Without a foundation in group self-regulation, management lacks the means to weld the miscellaneous collection of task groups into an effective team for the performance of the cycle as a whole. This situation is made more difficult by the fact that overall cycle responsibility is taken only at the level of undermanager, three steps in managerial rank from the coal face, since deputies and overmen have only shift responsibilities—for faces and seams respectively.

In the composite longwall system, the different organizational pattern removes the difficulties which stem from over-specialized work roles, segregated task groups, and lack of cohesion in the face team as a whole. There is a common paynote, based on a fixed rate, to which a bonus is added according to the amount of coal produced. All members of the team share equally. By this means they have a direct interest in the completion of the cycle and disagreements between small groups and arguments with management over amounts due are eliminated. The team undertakes full responsibility for allocating men to shifts and tasks and the methods devised give rise to multi-skilled roles and a ready experience of the consequences for others of neglected work. The task groups, although the same as on conventional longwalls with regard to the activities they carry out, are not segregated from each other but are interchangeable in membership. As soon as the scheduled work of a shift is completed, men spontaneously carry on with whatever activity is next in sequence so that subsequent groups gain time in hand against unpredictable interferences with the progress of the cycle, always to some extent likely in the underground environment. Such group regulation and continuity of face operations parallels the self-regulation and continuity characteristic of single place working. Under these conditions the deputy is liberated from detailed 'progress chasing' and seam management is able to concentrate on planning and maintaining the conditions and services which permit the cycle to proceed without disruption and work groups to aim for higher production targets.

Within the same longwall technology, composite organization was found to possess characteristics more conducive than the conventional to productive effectiveness, low cost, work satisfaction, good relations, and social health.

Further mechanization, whether partial or comprehensive, has major effects on work organization. There are fewer work roles—less narrowly specialized than those on semi-mechanized conventional faces—and fewer task groups—of a kind requiring an interlocking of the activities of different members. Each of these groups undertakes a larger part of the cycle than any group on a conventional longwall, so that the problem of integration is lessened. These direct consequences of increased mechanization tend to produce a work organization less complex than that on semi-mechanized conventional faces and one which has much in common with the characteristics of single place and composite systems. Acceptance of responsibility for the whole cycle, recognition of the dependence of one man or group on another and group self-regulation were characteristic of faces where higher mechanization had been introduced against the background of recent experience of the single place/composite tradition. These qualities were less in evidence where such experience was absent.

The conventional pattern of work organization on partially mechanized longwalls is to be regarded as a divergent development between single place working and the more comprehensively mechanized systems now being introduced. An alternative organizational form—the composite—has emerged and was operating more effectively than the conventional in the pits studied. Composite organization is consistent both with single place and with more highly mechanized systems. Not all features of single place organization—cavilling, for example—are appropriate to longwalls; nor do the composite longwalls seen necessarily represent an optimum organization for their particular level of mechanization; nevertheless, composite organization is better adapted than conventional to present longwall technology. In low seams the overall cycle group is much smaller than that found on long faces in higher seams and, for this reason, easier to handle on composite principles. In high seams, however, one side of a double unit can always be treated as a separate group, when it would not often be appreciably larger than the panel-wide groups studied in the present research.

How easily composite organization can be introduced in a colliery depends on the strength and recency of its single place tradition and, further, on whether this tradition has persisted in shortwall conveyor working. In low seam pits which are likely to continue at their present level of partial mechanization—and in which local tradition and experience are favourable—the introduction of a composite form of longwall organization would give promise of increased production, lower costs, better attendance, and improved relations, without added capital investment in equipment. In pits where there is no composite tradition, resistance to the introduction of composite working is likely to be considerable. Where this is the case and management still wishes to proceed, the resistances can be best overcome by full discussion and 'working through' of the attitudes and feelings involved among all concerned, both officials and workmen. These discussions need to be carried out in face-to-face groups which are not part of the formal negotiating machinery or executive structure. They require to be set up in a permissive atmosphere outside official channels, as 'working conferences' whose specific object is to permit attitude change to take place. Decision-making is left to subsequent meetings of the formal negotiating and executive groups.

Further mechanization gives the best opportunity to change work group organization. Unless, however, special steps are taken, such as those described above, inappropriate conventional patterns are likely to persist. Where higher mechanization is intended, advance introduction of composite methods at the existing level of mechanization may not only assist the transition but allow learning to take place which at a later stage will enable the group to operate the new equipment closer to its potential limit.

In changing over from conventional to composite organization, or to a more advanced technology, traditional methods of selecting and deploying teams do not lead to optimum results at the level either of the face or of the seam. New methods require to be worked out between management and lodge which allow full realization of the skill and leadership resources available in a seam population so that balanced teams are produced and key roles always occupied by specially competent individuals. This needs to be done in a way which gives as much scope to mutual choice by workmates as is consistent with these wider requirements.

When a new technology is introduced an inappropriate form of work organization—that associated with the technical trial of the machinery—tends to be carried over to subsequent operational units. A period of protection is needed to allow the explicit exploration and development of a satisfactory internal work organization. This period is considerably longer than that usually allowed for the acquisition of new skills.

Even when the effects on the social system are recognized—

albeit intuitively—there is often failure to communicate the lessons from the earlier to the later groups involved in a programme of reorganization. This is not because of a lack of facilities, but because the constructive outlook of management towards the dissemination of technical information and understanding does not extend to the socio-psychological dimension. There has, as yet, been no establishment of the idea of social learning through operational experiment.

For the full potential of highly mechanized continuous mining to be realized, a radical change is necessary from the practices and values of a man-centred to those of a machine-centred work culture. Implicit in a machine culture are a new value of time—because of the potentially high production rate of machines; a new basis for productivity—machine utilization; changes in underground organization—to permit continuous operations at the face; and a revaluation of traditional methods of manning—to ensure that key roles are filled by the best men and that teams are balanced. Some kind of 'unfreezing' of the existing situation is necessary before such a new culture can build up. Change to a higher level of technology, however limited, provides such an event but it cannot be of great value unless its potentiality for initiating social change is also recognized. It is the goodness of fit between the human work organization and the technological requirements that ultimately determines the efficiency of the whole system.

In traditional mining methods, control and regulation of work at the coal face were carried out autonomously by the working group, which developed customs of self-regulation, task continuity, and role-rotation appropriate to the underground situation. Seam officials provided services to the independently producing work places and, because of the slow tempo of production, had little difficulty in co-ordinating operations in the seam as a whole. In conventional longwall working where the face team is differentiated into a large number of segregated task groups, co-ordination and control have to be undertaken entirely by management. Such external control of facework operations involves officials in stressful and time-consuming bargaining and what has been called management through the wages system. They have in consequence less time to give to the maintenance and planning of seam services, which have become more complex as mechanization has increased. The more continuous production becomes, the higher the level of management required. The raised tempo of production calls for greater anticipation of future needs and for more rapid communication between the coal face and supporting seam services. The principal function of management in fully mechanized systems is to provide the conditions which permit faces of high potential productivity to achieve maximum output. This cannot be done if officials have to supervise in detail the work of groups at the face. A return to responsible autonomy by the face team offers the means by which officials can concentrate on providing effective communications within the seam, on anticipating the support required for operations at the face, and on giving attention to longer-term planning and development.

The feasibility of proceeding in this way depends on the assumption that relatively large primary work groups of up to 50 members are capable of sustained self-regulation and maintenance. The industry has been disinclined to believe this was possible and small group theory in social psychology has tended to support this viewpoint. The findings of this research are in the opposite direction. Self-regulating groups of this size were found not only to exist, but to persist in a steady state over considerable periods of time. Some of the conditions crucial for their successful emergence may be inferred from the contrasting outcomes on the panels described in detail in Sections Four and Five. Further studies in other contexts, as well as that of mining, would enable more of the relevant factors to be identified and more fully delineated models to be constructed of the type of system most likely to yield optimum results.

Appendices

APPENDIX I

Glossary of Mining Terms (North West Durham)

BACK-BYE The area and roadways between the coal face and the shaft bottom.
BACKSHIFT The second or middle shift of the day: varies from 9–10.30 a.m. until 4.30–6 p.m. in different pits.
BAIT STAND Mealtime during a shift. Time varies according to progress of the work, usually not more than 20 minutes.
BAND Layers of stone or shale found in some coal seams.
BANK The area around the top of the shaft.
BARGAIN An inclusive price agreement made between a set of men and management to complete a specified job, e.g. removing a small area of coal, driving a new roadway, winning out a new face, taking a caunch. Bargains are not subject to cavilling.
BORD AND PILLAR WORKING A system of mining in which interlacing roadways are driven at right angles into the seam, leaving small square or rectangular pillars of coal of from 30–50 yds side length, which are then wholly or partly extracted by a small group. (Also called Room & Pillar, Tub & Stall, Bord & Wall.)
BORDWAYS At right angles to the main plane of cleavage of the coal or cleat. In Durham bordways is roughly east and west.
BOX *See* Tension end.
BREAKING IN *See* Hewing.
BUTTOCK GETTERS Preparation-getting machines which take the coal at right angles to the face line on a buttock.
CAGE Lift for winding men or coal in a shaft.
CASTING Payment made to fillers when coal has to be shovelled more than an agreed distance to the conveyor belt or tubs.
CAUNCH Stone taken from the floor or roof to provide a roadway or gateway of adequate size. A caunch may be taken in two separate parts when it is divided into a fore caunch and a back caunch.
CAVILLING A system for the allocation of men and groups to work places within a seam, on a chance basis. Of special importance are the quarterly changes made at a ceremony between management and lodge.
CAVING Letting the roof collapse into the goaf in a controlled manner by removing supports.
CHOCKS Roof supports built of 4″ × 4″ × 18″ timber in a crosswise fashion, placed along the edge of the goaf. May include a steel quick-release device to

facilitate extraction. Are being superseded by fixed or adjustable all-steel chocks.

CLEAT Plane of natural cleavage of the coal.

COMPOSITE WORKING A system of organization in which all members of a team undertake all face tasks and share in the common paynote.

COMPREHENSIVE MECHANIZATION Preparation-getting machines in conjunction with flexible armoured (snaking) face conveyors and power-assisted stonework. The advancing of the conveyor and of 'walking chocks' are triggered by the passage of the cutter-loader in recent experiments in the introduction of automation. Comprehensive mechanization is still rare in low seams.

CONSIDERATION Payment made to a man or a team to make up earnings which are below an equitable level or as compensation for extra work or abnormal conditions.

CONVEYOR A means of moving material at the face or on roadways by an endless belt or scraper chain.

COUP An informal exchange of work place, task, or shift time.

CROP COAL Coal remaining on the floor after face has been undercut, caused by the cutterjib rising from the floor. Has to be taken up by picks. In higher, wet seams may be deliberately left to allow water to drain from the face into the goaf.

CUTTER-LOADERS *See* BUTTOCK GETTERS

CUTTER STALL A small area of coal flanking the mothergate which the cutter cannot reach and which is removed by hand. Usually this is made ahead of the face line and so facilitates cutter turning and provides easy access to the face when the cutter is parked at the mothergate end.

CUTTING The operation of undercutting coal with a mechanical cutter. The machine, which runs on electricity, employs two cuttermen.

CYCLE The complete sequence of face operations required to get coal.

DATAL (-WORKER) Day-wage workers employed in areas up to but not at the face, e.g. on the haulage.

DEPUTY A junior official responsible for safety precautions and mining operations in a face district.

DISTRICT An area of a seam for which a deputy is responsible. In N.W. Durham usually one face unit with its travelling, haulage, and air return roadways.

DOUBLE UNIT Two adjacent longwall faces—usually each of the same length—on either side of a main or mothergate. In N.W. Durham each face is usually 80–90 yds making 160–180 yds in all.

DRIFT An inclined roadway driven in stone either underground or from the surface to the workings.

DRILLER (-ING) Uses an electric or pneumatic twist drill to make shot-holes in the coal at 3′ 0″–6′ 0″ intervals. Shot-holes in the gateway caunches are usually 'put on' by the stonemen.

DUMMY GATE A small gate made on the face between the mothergate and tailgate for the purpose of getting stone to make strip packs for roof support (when goaf roof is supported and not allowed to cave).

E.M.S. Earnings per manshift.

ENDLESS ROPE HAULAGE A double track haulage system operated by an endless rope on to which tubs are clipped.

FACE RUN The time during which a coal-getting machine is moving along the face.

FACE SIGNAL A wire stretched along the face to control, directly or indirectly, the running of the face conveyor.

FACES On a double-unit longwall the faces may be in line, or one (leading) may be ahead of the other (lagging). The face at the end of the mothergate may be in line with the faces or in advance of both.

FILLER (-ING) Shovels the shot coal by hand onto a conveyor belt along the face.

FLAT In single place workings the area served by one or more putters.

FLIGHT-LOADERS The ordinary coal-cutting machine may be fitted with an automatic 'gummer' which throws out the small coal from the undercut onto the face conveyor belt. After the coal has been undercut (and fired down if necessary) some of the cutting picks are replaced with paddles or 'flights'. The machine travels back along the face, the flights loading the broken coal onto the face conveyor. This method is a simple form of power-loading.

FLIT Move or transport (applied to coal cutting equipment).

FOREOVERMAN Official responsible for the working in a seam during the first (fore) shift of the day. Next in seniority to the undermanager.

FORE POLING A method of roof support in which horizontal bars are supported by a cantilever arrangement in front of the foremost supports on the face or in a gate.

FORESHIFT The first shift of the day: starting time varies from pit to pit and from seam to seam: 12.00 p.m.–4.30 a.m. lasting until 7.30 a.m.–10.00 a.m.

FRONTAL GETTERS Preparation-getting machines which shear off the coal in thin slices along the whole length of the face.

GATEWAY LONGWALL A continuous coal face served by gateways (in Durham about 12 yds apart). A small group works in each gateway down which the coal is removed by tubs.

GEARHEAD The motor drive, switch gear, and unloading device of a conveyor belt. Face conveyor gearheads may stand in the mothergate or in a caunch at the side of the gateway.

GETTING The mining or extraction of coal from the seam.

GOAF (GOB) The waste area behind the face from which the coal has been extracted. May be wholly or partly packed with stone to support the roof, or the roof may be allowed to collapse (see CAVING).

GUMMER *See* SCUFFLER

GUMMINGS *See* KIRVINGS

HEADING A roadway being driven, generally in the coal seam, in a headways direction.

HEADWAYS Parallel with the main cleavage plane or cleat of the coal. In Durham this runs north and south.

HEWER (-ING) May use a hand-pick but usually a pneumatic ('windy') pick to win coal. Task consists of 'breaking in' or making a 'nicking', digging out the coal, and filling onto a conveyor belt or into tubs.
HITCH A minor geological fault or roll in the coal seam.
IN-BYE In a direction towards the working face and away from the shaft. Reverse of out-bye.
INTAKE Any roadway underground through which fresh air is conducted to the working face.
JIBBING IN The first operation of cutting—with the cutting machine flush and parallel to the face—bringing the cutterjib at right angles into the coal to start the undercutting.
KIRVINGS Small coal lying in the undercut made by cutting machine jib.
KIST The meeting place in a district where workmen assemble. (Strictly the box in which the deputy keeps his papers, etc.)
LANDING Area where tubs are assembled and connected to the seam haulage system.
LOADING POINT Where coal is transferred from a mothergate or trunk belt conveyor into tubs.
LONGWALL MINING A system of mining on straight faces 80 yds or more in length.
MACHINE END *See* GEARHEAD
MAIN AND TAIL HAULAGE A single track rope haulage system in which the rope is fastened to the first and last tubs of a set.
MAKE UP Payment made by management (deputy and overman) when earnings from agreement fall below a seam 'norm'.
MARROW A work-mate or partner. In single place working from two to six men may 'marrow up' to form a set which pools and shares earnings equally. Marrows may work on different shifts although always in the same work place.
MASTERSHIFTER Official responsible for the working of a seam during the third (night) shift of the day.
MEETING STATION *See* KIST
MOTHERGATE The main roadway to a coal face up which men travel, air, power, and supplies pass, and down which coal from the face travels on a conveyor belt.
MULTI-JIB CUTTERS Instead of the single undercutting jib, two or three straight jibs cut horizons into the coal, a curved jib makes a 'vertical' cut at the back, a 'mushroom' jib may break down the top coal. With such machines in soft coal, shotfiring is unnecessary.
MULTIPLE TASK MACHINES Any machine which mechanizes more than one task of the cycle.
NEUK (-END) The tailgate corner of a face behind the face conveyor tension end.
NIGHTSHIFT The third shift of the day: starting time varies from 4.30 p.m.–12.00 p.m.
NOGS The wooden blocks used to build chocks. Also used as wedges for supports and to hold up the undercut.

NOTE Itemized list of payments due to a man or a set of men for work done during a week.

O.M.S. Output (usually tons) per manshift. Interpretation depends on the basis for calculating manshifts, e.g. *face* o.m.s. is based on manshifts at the face; *seam* o.m.s. on piecework and bargain work manshifts in the seam; *overall* (pit) o.m.s. on all manshifts underground, including datal labour.

OUT-BYE In a direction away from the working face towards the shaft. Reverse of in-bye.

OVERMAN An official responsible for the working of a seam during a shift.

PACK Retaining walls on either side of a gateway or road, built in the goaf of large stones and packed with rubble. Usually from 3–6 yds wide in low seams.

PANEL *See* DOUBLE UNIT

'PENCIL' *See* CONSIDERATION

PIECEWORK (-ER) Workers not paid on a day-wage, i.e. paid by results.

PLOUGHS Vertical and horizontal cutting blades which remove a thin slice of coal from the face and guide it onto an 'armoured' face conveyor on which the machine runs (other forms of cutter-loader may travel also on such a heavy scraper chain type of conveyor).

POWER-PULLING Normal pulling of the face conveyor requires the belt to be broken into convenient lengths, rolled up, and pulled in the new track. In power-pulling, each half of the face belt is pulled around a prop at one end of the face by means of a rope attached to a small electric winch.

PREPARATION-GETTING MACHINES Any machine which combines the preparation and getting phases of the cycle.

PROPS Timber or steel supports for the roof. Steel props may be adjustable.

PUFFLER A member of a shift or task group informally recognized by the group as a spokesman. Takes up matters with management on behalf of the team. (In some pits, used as a term of abuse.)

PULLER (-ING) Moves forward the face conveyor equipment and withdraws and rebuilds chocks and face supports as the coal face advances.

PUTTER (-ING) In single place systems a man, usually a young lad, who conveys tubs between the workings and a flat. He may have a pony.

RAMBLE Loose stone which overlays the coal and falls when coal is removed.

RETURN (AIR) Any roadway through which air returns to the surface from the working faces.

RIB AND STALL WORKING A gateway with a face to one side about 6 yds in length (the stall) is advanced for up to 50 yds. Another face (the rib) is then retreated down the other side of the gateway already made. Worked by a small group of men, usually one on shift at a time, doing all operations.

RIBSIDE Exposed coal flanking one side of a roadway.

ROLL A bulge in the floor or roof or both, causing lessening of the coal height.

SCRAPER-BOXES In this system a thin web of coal is removed by the vertical cutting edges of a series of boxes which move to and fro across the face. The coal so peeled is scraped into the boxes, and transferred from one to another until it reaches the gate conveyor.

SCRAPER-PACKER A machine-operated drag-line and scoop, for removing stone from the gateway and packing it in the goaf. Also called a slusher.
SCUFFLER Man working with a cutting team responsible for cleaning out the kirvings from the undercut coal and placing nogs or sprags to support the coal. Known also as GUMMER or DUFFER.
SEAM A bed of coal. Seams less than 3′ 0″ in height are regarded as 'low'. Seams down to 18″ in height are commonly worked, seams lower still, more exceptionally.
SET (a) A number of tubs (usually of 10 cwt capacity) coupled together. Up to about 60 or 70 form a 'set'.
(b) A facework team or group.
SHIFT WORK Work, not included in a particular agreement, which is done and paid for at a fixed rate.
SHORT WALL WORKING An enlarged version of rib and stall working with faces of up to 50 yds in length. Gateways are made at both ends of the face.
SHOTFIRER (-ING) An official whose sole responsibility is to charge, stem, and fire down undercut coal in preparation for filling.
SINGLE PLACE WORKING Systems in which one man works in each place on a shift, doing all coal hewing and stonework tasks as necessary to advance the face. (Includes Bord & Pillar, Rib & Stall, Stepwise and Gateway Longwall systems.)
SINGLE TASK MACHINES Any machine which mechanizes one task of the cycle.
SINGLE UNIT A longwall face with a mothergate and tailgate but no second face across the mothergate. In N.W. Durham 70–100 yds, usually 80 yds.
SLUSHER *See* SCRAPER-PACKER
SPRAGS Pieces of wood or steel used to support undercut coal before it is broken down by shotfiring.
STAPLE A spiral chute down which coal passes from one seam to the haulage system of a lower seam. Can act as a storage bunker in case of haulage breakdown.
STEEL MONEY Payment made to faceworkers who have to handle steel straps instead of timber planks.
STEPWISE LONGWALL A variant of gateway longwall in which faces 12–40 yds in length are 'stepped', each face being 4–6 yds in advance of the adjacent one.
STINT The length of face, fixed by agreement, to be cleared by each man on a longwall filling shift.
STONEMEN (-WORK) Face (gateway) stonemen make the mothergate and tailgate to the face, shovel the shot stone from the caunches into the goaf, build packs, and erect gateway supports.
STRAPS Corrugated steel bars about 4″ wide and 6′ 0″, 9′ 0″ or 12′ 0″ in length, held against the roof by props. Are replacing timber planks. Usually placed at right angles to the face and 3′ 0″–4′ 0″ apart.
STRATA CONTROL The system of props, straps, chocks, and packs used to maintain control of the roof and floor strata.

STRIP PACKS *See* DUMMY GATE
SYLVESTER A simple lever and chain device used for withdrawing tight props or chocks or moving heavy equipment.
TAILGATE A subsidiary roadway at the end of the face away from the mothergate down which air passes. May be used as a supply road.
TEAM CAPTAIN *See* PUFFLER
TENSION END (or BOX) A drum or drums around which a conveyor belt passes, with some device for applying tension to the belt.
TOKEN Disc placed in loaded tubs to indicate to which man or groups it should be credited.
TRACKING THROUGH Clearing a path along the whole length of the face through the newly fired-down coal. The first operation of filling.
TUMBLER END *See* TENSION END
TURN ROUND The time involved in turning or flitting a coal-getting machine.
UNDERMANAGER Senior official responsible for the working of a seam or number of seams. Each separate ventilation system is required by law to have at least one undermanager.
WAITING-ON Time paid for at a fixed rate when normal filling work cannot be done because of events outside the men's control or when shift work is not available.
WASTE *See* GOAF
WEB A slice or strip of the coal seam taken off along the length of a coal face.
WET MONEY Payment made to faceworkers who have to work in wet conditions.
WINNING The getting and loading of coal.

APPENDIX II

Composite Development on Bramwell Faces after Reorganization[1]

Ratings of degrees of compositeness on scale (range 0–12) as described in Tables 28 *and* 29 *(Chapter XXV); completeness and manpower indices as in* Table 31 *(Chapter XXVI).*
Faces arranged in order of increasing degrees of compositeness.

FACE *F*: COMPOSITE RATING (4); COMPLETENESS (77); MANPOWER (112)

This face, which averaged four cycles a week, started nine weeks after reorganization. It had dry working conditions, but was subject to minor hitches and rolls. Originally 90 yds, its length was reduced three times in 13 weeks to avoid an old working and a fault. Finally, it became a 65-yd face, but the 24-man team was not reduced as two men were permanent absentees. None of the team had composite experience; half were single place workers and the rest had been on conventional cutting or hewing faces. Two-thirds had been on the spare cavil and had never worked together, having been employed as substitutes on other faces. The work organization adopted was 'one man—one job', with permanent cuttermen, driller, pullers, and stonemen. One of the two scuffling roles was taken permanently by one man and the other rotated on a weekly basis among the fillers.

During the first three months a great deal of tension centred on the cutting. The floor was hard and caused the jib to crop the coal. According to the fillers and pullers much of this was due to the bad technique of the senior cutterman who had not been employed on cutting for many years. The two cuttermen could not agree, and when the scuffler was trained the No. 2 cutterman went on the filling shift. In the end, the team took the extreme step of asking management to remove the No. 1 man, the 'last straw' having been the cutting of the face during a week-end by a deputy who made a first-class job of it. When the man was removed, the former No. 2 cutterman returned.

There were also disagreements about how the work was to be shared out. Each shift, in the words of the deputy, 'seemed to be for itself'. As there were

[1] Source paper: 14.

only two qualified cuttermen the filling and stonework shifts had no one who could complete or start the cutting when the state of the cycle required such activities (*Table 37*). The filling shift was short of qualified pullers, particularly later when the only qualified filler-puller became a permanent scuffler. There was little scope, therefore, for the fillers to push the cycle ahead. On the cutting and stonework shifts, however, some task continuity was carried out; cuttermen and scufflers would track through and mothergate stonemen would pull up the gearhead.

On the filling shift the attitudes towards cycle completion were more like those on conventional faces, with scapegoating of the previous shift and little account taken of the requirements of pullers and stonemen. The fillers were not a well-knit group; only one had experience of cutting longwalls, though another had worked on a hewing longwall; five had been hewers in single places. The pullers and stonemen had a more composite attitude; of the eleven men on this shift at least eight had worked on conveyor longwalls and the previous marrow groups had not been broken up to the same extent as on the filling shift.

TABLE 37 SKILL RESOURCES ON FACE *F*

Work Role	*No. of Workers Qualified in* *C*	*D*	*F*	*P*	*S*	*No. Required in role*
Cutting/Scuffling	2	3	3	3	3	3
Drilling		1	1		1	1
Filling			7	1	7	7
Pulling		2	2	4	4	4
Stonework		3	6	4	7	7
Total Team	2	9	19	12	22	22[1]

[1] Two permanent absentees.

FACE *B*: COMPOSITE RATING (6); COMPLETENESS (76); MANPOWER (103)

During its nine weeks of life, this face had trouble-free working conditions but the large size of the mothergate caunches (taken because the gate was being kept open for another face) caused delay in the stonework. An average of four cycles a week was achieved. The team was composed of roughly equal numbers having conventional longwall and single place experience. Work roles developed in the pattern of one task on two shifts. There were permanent cuttermen, pullers, and stonemen who changed their shift times week and week about. The fillers, however, rotated in pairs for fortnightly tours on to scuffling; and, though drilling was a permanent job at first, after a month it also began to rotate among the fillers. There was no couping or taking up of absentee vacancies on the backshift, though within the shift there was movement between activity groups; for example, a mothergate stoneman would take on an absentee puller's role.

Cycle disorganization came about when the cuttermen were prevented from cutting the face right through, as when mothergate stonework, through the size of the caunches, was unfinished and the face and gate belts were unable to run. There were only two qualified cuttermen so that when reserves were unavailable from outside the group, men still under training had to complete the cut or the cycle came to a halt.

Though the work roles they developed were in the conventional tradition, they worked in the composite manner, and when the cycle was out of phase shift groups redeployed to undertake to the best of their ability whatever roles were necessary (*Table 38*). The achievement of a steady cycle flow, however, was hampered by the absence of an internal reserve of cuttermen who could work on the filling shift and cope with cycle lag arising on the cutting shift.

TABLE 38 QUALIFICATIONS AND EXPERIENCE AVAILABLE ON FACE *B*

Work Roles	*No. in Same Role before Reorganization*	*No. of Workers Qualified in* C	D	F	P	S	*No. Required in Role*
Cutters	2	*2*	*2*	*2*	*2*	*2*	2
Driller	0		*0*	1		1	1
Fillers	8			*10*	3	10	10
Pullers	0			5	*4*	5	5
Stonemen	3		6	7	5	*8*	8
Total	13	2	8	25	14	26	26[1]

[1] One permanent absentee.

FACE *G*: COMPOSITE RATING (9); COMPLETENESS (95); MANPOWER (103)

This face, which started eleven weeks after reorganization, had the best working conditions in the seam and completed five cycles per week. Of the 23 men, 19 had worked together since reorganization on faces *B* and *S*; the other four had been together on *C*, which had developed a form of shift rotation; and 2 of these men had previous composite experience. Work roles were of the one-task—two-shift variety, similar to those developed on *B*, 16 men occupying the same roles on both faces. Two of the men from *C* took on the cutting, while the composite men, also qualified cuttermen, split up, one going on to filling and the other on to stonework. This face, therefore, had cuttermen on all shifts—a quite deliberate deployment (*Table 39*). Only one scuffler was used whose role rotated among the fillers on a fortnightly basis.

TABLE 39 SKILL RESOURCES ON FACE G

Work Role	*No. of Workers Qualified in* C	D	F	P	S	*No. Required in Role*
Cutting/Scuffling	*2*	2	2	1	2	2
Drilling		*0*	1		1	1
Filling	1	1	*8*	3	8	8
Pulling		1	4	*3*	4	4
Stonework	1	5	8	5	*8*	8
Total Team	4	9	23	12	23	23[1]

[1] One permanent absentee.

The men had gained some experience of composite working under the generally favourable conditions of their previous faces. Task continuity was practised on all shifts in much the same way as in the Manley, and the physical appearance of the faces was similar. Three features found on Manley longwalls emerged—complementary shift-sharing, role exchanges lasting for two weeks, and movement into absentee vacancies on the back-shift. These occurrences were not frequent (fifteen in as many weeks), but they suggest that Manley customs were becoming known in the Bramwell.

FACE *C*: COMPOSITE RATING (11); COMPLETENESS (75), ADJUSTED (93); MANPOWER (98)

This lasted eleven weeks and had good conditions but was equipped with the old type of Blackett scraper-chain face conveyor which made for more arduous filling and pulling. It achieved four cycles per week. Before reorganization two-thirds of the team had worked on conventional and one-third on composite cutting faces. It had no shortage of cuttermen, six being qualified. Two of these undertook responsibility for cutting while the others distributed themselves on the filling and stonework shifts to cope with lag or to prepare the machine. The two scuffling roles were occupied by nine different men, three of whom spent half their time at this task.

Only one man carried out one task on two shifts throughout the whole period. All the others developed multi-task—multi-shift roles, those whose previous experience had been on conventional cutting longwalls 'converting' to composite types of role (*Table 40*).

The team followed the composite shortwall custom of agreeing among themselves at the end of each week the shifts they would work during the next. The result was disparity in the amount of time on the backshift: in the eleven weeks, six men had only one or two weeks, whereas twelve men had five weeks or more. This situation resembles closely that found in the first phase of No. 1 Panel (East) in the Manley, which resulted in the introduction of complementary shift-sharing. Had the team not been aware of its impending

TABLE 40 TASK-SHIFT ROLES ON FACE *C*

Role	*Previous Experience* Conventional	*Composite*	*Total*
1 Task —2 Shifts	1	–	1
2 Tasks—2 Shifts	1	1	2
2 Tasks—3 Shifts	4	2	6
3 Tasks—2 Shifts	1	–	1
3 Tasks—3 Shifts	8	4	12
Total Team	15	7	22

disbandment, a similar development might have occurred on *C*, for two men had already begun to share a scuffling and a filling role. The most interesting feature was the extent to which a large group of men with conventional experience readily adopted the customs of a smaller composite group. Being dependent on the few composite men for the crucial tasks of cutting and pulling, they were prepared to give a trial to an alternative method.

FACE *D*: COMPOSITE RATING (12); WITH BLACKETT CONVEYOR, COMPLETENESS (42), MANPOWER (74); WITH BELT CONVEYOR, (51)–(72)

Face *D* presents a complete contrast to the other faces. It had experienced a number of hitches and was so wet ('more coal is washed off the belt than gets into the tubs') that cutting and stonework were frequently held up. Two or three cycles only were completed per week.

The men had worked together for seven years as a composite team gaining a high reputation. All were qualified in filling, pulling, and stonework and two-thirds in cutting.

The original allocations on *D* were in terms of a small cutting shift along with filling and pulling-stonework shifts of approximately equal size. This arrangement was not appropriate with a cycle completely out of phase, and the team gradually altered its deployment until the three shifts became equal.

The rotation system was a continuation of that which the team had developed over the years (*Table 41* (*a*) and (*b*)).

There were five rotation groups, each working a different sequence. Three of the groups (a, b, and c) spent equal amounts of time on all three shifts with each group always on a different shift. The two other groups followed the complementary principle as regards 'bad' shifts, one (d) taking the foreshift and the other (e) the night. These patterns were adaptations of shortwall practice in which three groups of equal size worked the straight sequence of fore, back, and night. They guaranteed to all the same amount of backshift, but allowed the exercise of preference with regard to the other shifts. Instead of three separate shift-rotating groups whose members worked only with each other, two-thirds of the team (groups b–e) worked on the same shift with other

TABLE 41 SHIFT ROTATION SYSTEM: FACE *D*

(a)

Rotation Group	*Shift Sequence*	*No. of Men in Group*	*No. Qualified to Cut*
a	Fore, Back, Night	6	6
b	Back, Night, Fore	3	1
c	Night, Fore, Back	3	0
d	Back, Fore, Fore	3	3
e	Night, Night, Back	3	3

(b)

Shift	*Rotation Groups Constituting Shift* *Week 1*	*Week 2*	*Week 3*
Fore	a[1]	c, e	b, e[1]
Back	b, e	a	c, d
Night	c, d	b, d[1]	a

[1] Cutting alternated between fore- and nightshifts every week.

members. Only group (a) followed the strict shortwall model of providing a complete shift group on all three shifts. Since groups (b) and (c) were short of cuttermen the system ensured that there were never less than three men qualified to cut on any shift.

Individual coups were arranged and when men were absent their good shifts were taken by other team members, substitutes going on to the bad. Since the number of cycles completed each week varied, the tasks to be carried out on any particular shift were unpredictable. Task preferences, therefore, played no part in the rotation system, though they were considered in intra-shift deployment. The men were so familiar with each other's skills and preferences that, in the words of one deputy, 'no one gives any directions, everyone knows just what to do and how to do it'. This behaviour is identical with that found in the teams on the Manley longwalls, many of whose members had worked on composite shortwalls in the Bramwell alongside team *D*.

APPENDIX III

A Model of Different Stages in the Development of Composite Organization on a Conventional Cutting Longwall[1]

This appendix sets out a model for stages in the development of composite organization. The calculations are based on optimum conditions, and in practice would have value as an aim rather than as an expectation of actual performance. Their error, however, should not be greater than 10 per cent.

TABLE 42 HOURS PER CYCLE FOR MAIN TASKS AND NON-CYCLE ACTIVITY, AND ESTIMATED NORMAL TIME REQUIRED FOR ALL MAIN TASKS

1 *Main Task Group*	2 *No. of Scheduled Manshifts*[1] *per Cycle*	3 *Hours for Main Task*	4 *Hours for Non-Cycle Activity*	5 *Estimated Hours for Normal Interference*	6 *Estimated Gross Hours for Main Task* (3+5)
Cuttermen	4	13·3	10·7	1·1	14·4
Scufflers	2	6·7	5·3	0·5	7·3
Fillers	12	56·2	15·8	0·8	57·0
Driller	1	3·3	2·7	0·1	3·4
Hewers	1·5	7·5	1·5	0·2	7·7
Pullers	6	21·4	14·6	1·5	22·8
Stonemen	11	62·5	3·6	0·4	62·9
Totals	37·5	170·9	54·1	4·5	175·5

[1] Effective manshift = 6 face hours.

The figures in *Table 42* have been calculated from the returns over one quarter for the conventional face discussed in Chapter XIII when conditions were exceptionally constant. They exclude overtime and reinforcement, and the figures for filling and pulling are limited to the twenty cycles in which the fillers completed their main task and the pullers had no coal to take. An

[1] Source paper: 4.

allowance for normal interference of 10 per cent of the figures for non-cycle activity has been used. For the driller and fillers this has been reduced to five per cent, as for these groups half the interference they suffered originated outside the face district in the seam transport system and, in the estimate, external interference is presumed to be eliminated.

TABLE 43 MANPOWER REQUIRED AND MANPOWER SAVED PER CYCLE AT DIFFERENT STAGES OF COMPOSITE ORGANIZATION

(External interference is presumed to be negligible)

Stage	*Groups Combined*	*Hours for Combined Main Tasks*	*Manshifts*[1] *for Combined Main Tasks*	*Hours of Non-Cycle Activity Saved*	*Manshifts Saved*
1	Scufflers and Fillers	64·2	10·7	19·8	3·3
2	Cutters, Scufflers, and Fillers	79·6	13·1	29·4	4·9
3	Cutters, Scufflers, Fillers, Pullers, and Driller: *Face Team*	104·8	17·5	45·2	7·5
	Stonemen and Hewers: *Gate Team*	70·5	11·8	4·5	0·8
4	All Groups Combined: *Full Composite*	175·4	29·2	49·7	8·3

[1] Effective manshift = 6 face hours.

Table 43 sets out the estimated manhours and manshifts required for the main tasks of different groups and the estimated saving that would result from successive approximations to complete composite working. Estimates for different stages are given because it is likely that in an old-established conventional system it would be neither practicable nor desirable to attempt a complete change at one stroke even with a significant 'unfreezing' event. The process may also become arrested at different points.

The *first stage* would be the taking over of scuffling by the filling team—already common practice in a number of pits. With continuity between scuffling and filling there should be a saving of the 3·3 manshifts now lost to the cycle on the scuffling task, mainly from inappropriate manning. This, together with the saving on the filling shift from the assumed elimination of external interference and a 90 per cent reduction of internal interference, would mean a total saving from this combination of over three manshifts.

The obvious *second stage* would be to combine the cutting roles with this enlarged group. The cumulative saving would now be 4·9 manshifts, mainly through task continuity on the cutting shift. A second advantage of such a combination would be that instead of work in the face area being done by three shift groups, two of whom (pullers and cuttermen) never meet in the pit and may not even be acquainted outside, it would now be done by only two groups—a cutting/filling group and the pullers, who would meet each other at shift changes. Such a two-way relationship of men sharing the same work place would make for easier relationships than the former three-way 'open triangle' pattern. A third, and by no means unimportant, result would be that men doing cutting would not be condemned always to work on either the late nightshift or the foreshift. A simple system of task rotation would give them a chance to share the backshift—the only 'good' shift.

In the *third stage* the pulling and drilling roles would be absorbed into the cutting/filling combination to make a single *face team* responsible for all the tasks done in the face area. Thus for the men on face tasks the conditions for reaping the advantages of task continuity, cycle consciousness, and minimal social tensions would be secured, a situation in which all social forces would be affecting behaviour in a way beneficial to cycle completion and the realization of output potential. Cumulative saving would now be 7·5 manshifts.

A second part of this third stage would be the combination of all the men working in the gate areas—the four groups of stonemen and the hewers in the mothergate—into one *gate team*. The combination of the tailgate and mothergate stonemen is a natural one. They all do the same task and have similar levels of pay. Their combining into one team would give them the advantages of mobility between gates in order to meet varying conditions, and of greater internal specialization than is possible in very small independent groups of two and three. The hewers, though getting coal, have less in common with fillers with whom they are on shift than with the forecaunch stonemen with whom they share a work place and a common task, namely, advancing the mothergate. Also, the hewing task requires only one and a half manshifts per cycle. In the estimate this has been allowed for but, in practice, if a gate team of twelve were to detail two men for hewing, the extra half shift could be used to prepare the caunches for stonework. This, together with the saving already mentioned arising from working as a larger team, would allow the odd half shift to be saved. It might also be found that more than two men from the gate team could work while the face operations were going on, as it is not unusual for tailgate stonework to be going on at this time. If this were done, half the members could be on shifts other than the nightshift, which now is the only shift for stonemen, and a rotation system would become possible.

The stonework shift is a heavy and full shift and contains very little non-cycle activity. Approximately another manshift would be saved by amalgamation with the face team. In our observation of full composite teams, however, we have seen how spare time from a smoothly running cycle can be put into stonework. This allows the stonemen to get sufficiently ahead to prepare for and even to commence cutting, in order to get the next cycle away to a good start. With the stonework roles included in the full face group, a much slower rotation through the disliked nightshift also becomes possible, while cycle regulation attains much greater flexibility, with the stonemen helping the pullers when the chocks are difficult.

The changes these models suggest would not be achieved without arousing resistance. This would probably be especially true of the pullers. Should the combination of stage one and stage two be successful in channelling non-cycle activity into filling time, the lag normally inherited by pullers would be diminished. This would remove the foundation on which their favoured position is built, with its special payments resulting from cycle dysfunction.

Provided, however, the resistances were not strong enough to prevent the amalgamation of the other groups, the higher pay this increased productivity would yield all round and the lowering of their own pay as cycle dysfunction was eliminated would make the pullers more inclined to become part of the larger group.

Special Bibliography

SOURCE PAPERS

1. HERBST, P. G., Study of an Autonomous Working Group (pp. 80). T.I.H.R. Rep. 423, 1958.
2. HIGGIN, G. W., A Conventional Cutting Longwall I: The Technical System (pp. 8). T.I.H.R. Rep. 466, 1957.
3. HIGGIN, G. W., Face Team Organization and Maintaining Production (pp. 12). T.I.H.R. Rep. 469a, 1957.
4. HIGGIN, G. W., System Potential and System Change (pp. 12). T.I.H.R. Rep. 469, 1957.
5. HIGGIN, G. W., A Conventional Cutting Longwall II: Task Roles and Social Relationships (pp. 11). T.I.H.R. Rep. 486, 1958.
6. HIGGIN, G. W., Social Factors in the Change to Higher Mechanization in Longwall Working (pp. 26). T.I.H.R. Rep. 502, 1958.
7. HIGGIN, G. W., MURRAY, H., POLLOCK, A. B., & TRIST, E. L., Composite Cutting Longwalls V: The Development of Task-Shift Rotation Systems (pp. 37). T.I.H.R. Rep. 420b, 1957.
8. MURRAY, H., Composite Cutting Longwalls II: Work Organization (pp. 20). T.I.H.R. Rep. 463, 1957.
9. MURRAY, H., Composite Cutting Longwalls III: The Production Record of Two Double Unit Longwalls (pp. 12). T.I.H.R. Rep. 464, 1957.
10. MURRAY, H., Composite Cutting Longwalls IV: The Assessment of Performance (pp. 22). T.I.H.R. Rep. 465, 1957.
11. MURRAY, H., Composite Cutting Longwalls VI: Work Roles and Activity Groups (pp. 25). T.I.H.R. Rep. 473, 1957.
12. MURRAY, H., Composite Cutting Longwalls VII: Social Adaptation to Increased Work Stress (pp. 20). T.I.H.R. Rep. 472, 1957.
13. MURRAY, H., Composite Cutting Longwalls VIII: Regulation of Cycle Progress (pp. 23). T.I.H.R. Rep. 482, 1957.
14. MURRAY, H., Changing from Conventional to Composite Organization (pp. 30). T.I.H.R. Rep. 484, 1957.
15. POLLOCK, A. B., Single Place Working (pp. 20). T.I.H.R. Rep. 420a, 1957.
16. POLLOCK, A. B., The Conventional Longwall Hewing System (pp. 26). T.I.H.R. Rep. 504, 1958.
17. POLLOCK, A. B., A Composite Hewing Longwall Unit (pp. 26). T.I.H.R. Rep. 504a, 1958.
18. TRIST, E. L. & BAMFORTH, K. W., Some Social and Psychological Consequences of the Longwall Method of Coal Getting (pp. 38). Hum. Relat., Vol. 4, 1951.

19. TRIST, E. L., Some Observations on the Machine Face as a Socio-Technical System, A Report to the Area General Manager, No. 1 Area, East Midlands Division (pp. 19). T.I.H.R. Rep. 341, 1953.
20. TRIST, E. L., Progress Report (pp. 25). T.I.H.R. Rep. 434, 1956.
21. TRIST, E. L., Composite Cutting Longwalls I: Their Emergence and General Characteristics (pp. 10). T.I.H.R. Rep. 420, 1957.
22. TRIST, E. L., The Problem of the Appraisal of Work Systems (pp. 5). T.I.H.R. Rep. 434a, 1957.
23. WILSON, A. T. M. & TRIST, E. L., The Bolsover System of Continuous Mining, A Report to the Chairman, East Midlands Division, N.C.B. (pp. 26). T.I.H.R. Rep. 290, 1951.
24. WILSON, A. T. M., Two Contrasted Coal Mining Systems in Some Aspects of Social Process (pp. 24). Journal of Social Issues, Supplement No. 5, 1951.

General Bibliography[1]

ANGLO-AMERICAN COUNCIL ON PRODUCTIVITY. *Productivity Team Report on Coal.* London, New York: Anglo-American Council on Productivity, 1951.

BALDAMUS, W. *Efficiency and Effort.* London: Tavistock Publications, 1961.

BALDAMUS, W. Type of Work and Motivation. *Brit. J. Sociol.*, 1951, **2**, 44.

BALDWIN, G. B. *Beyond Nationalization.* Cambridge, Mass.: Harvard University Press, 1955; London: Oxford Univ. Press.

BELL, D. *Work and Its Discontents.* Boston: Beacon Press, 1956.

BERTALANFFY, L. von. The Theory of Open Systems in Physics and Biology. *Science*, 1950, **3**, 23.

BION, W. R. Experiences in Groups: V. *Hum. Relat.*, 1950, **3**, 3.

BION, W. R., *Experiences in Groups.* London: Tavistock Publications, 1961. New York: Basic Books, 1962.

BLAU, P. *The Dynamics of Bureaucratic Structure: A Study of Interpersonal Relations in Two Government Agencies.* Chicago: University of Chicago Press, 1955.

BRADFORD, L. *Explorations in Human Relations Training: An Assessment of Experience, 1947–52.* Washington, D.C.: National Training Laboratories, 1953.

BROWN, W. B. D. *Exploration in Management.* London: Heinemann; New York: Wiley, 1960.

BRYAN, Sir Andrew. Economical Use of Mining Manpower. *Iron and Coal Trades Review*, Nov., 1954.

BULLMAN, H. F. & REDMAYNE, R. A. S. *Colliery Working & Management* London: Crosby Lockwood, 1906 (2nd ed.)

BURNS, T. & STALKER, G. *The Management of Innovation.* London: Tavistock Publications, 1961, Chicago: Quadrangle, 1962.

CARTWRIGHT, D. & HARARY, F. Structural Balance: a generalization of Heider's Theory. *Psychol. Rev.*, 1956, **63**, 277.

CARTWRIGHT, D. & ZANDER, A. (Eds.) *Group Dynamics: Research and Theory.* (2nd ed.) Evanston, Ill.: Row, Peterson, 1960; London: Tavistock Publications, 1960.

COOK, P. H. *The Productivity Team Technique.* London: Tavistock Institute of Human Relations, 1951.

DENNIS, N., HENRIQUES, F. & SLAUGHTER, C. *Coal is Our Life.* London: Eyre & Spottiswoode, 1956.

[1] A comprehensive bibliography of the literature relevant to the development of a general theory of socio-technical systems has been prepared by Emery in *Characteristics of Socio-Technical Systems* (London: Tavistock Institute, Doc. No. 527, 1959). Only works cited in the text together with certain others of general background significance are listed.

DEPARTMENT OF SCIENTIFIC AND INDUSTRIAL RESEARCH. *Automation: The Technical Trends and Their Impact on Management and Labour*. London: H.M.S.O., 1956.

DICKSON, D. E. The Morbid Miner. *Edin. Med. J.*, 1936, p. 696.

EMERY, F. E. Characteristics of Socio-Technical Systems. London: Tavistock Institute, Doc. No. 527, 1959.

EMERY, F. E. & OESER, O. A. *Information, Decision and Action*. London: Cambridge Univ. Press, 1958.

EMERY, F. E. & TRIST, E. L. *Socio-Technical Systems*. Proc. 6th Annual International Meeting of the Institute of Management Sciences. London, New York, Paris, Los Angeles: Pergamon Press, 1960. *Management Sciences Models & Techniques Vol. II*.

FESTINGER, L. *A Theory of Cognitive Dissonance*, Evanston (Ill.): Row, Peterson & Co., 1957.

FORTES, M. (Ed.) *African Political Systems*. London: Oxford Univ. Press, 1940.

FYNES, R. *The Miners of Northumberland & Durham*. Sunderland: Summerbell, 1873.

GOULDNER, A. W. *Patterns of Industrial Bureaucracy*. Glencoe: Free Press, 1954; London: Routledge & Kegan Paul, 1955.

HALLIDAY, J. L. *Psychosocial Medicine*. London: Heinemann, 1948.

HEIDER, F. Attitudes and Cognitive Organization, *J. Psychol.*, 1946, **21**, 107.

HEIDER, F. *The Psychology of Interpersonal Relations*. London: Chapman & Hall, 1958.

HERBST, P. G. *Autonomous Group Functioning*. London: Tavistock Publications, 1962.

HERBST, P. G. Measurement of Behaviour Structures by Means of Input-Output Data. *Hum. Relat.*, 1957, **10**, 335.

HERBST, P. G. A Theory of Simple Behaviour Systems. I. *Hum. Relat.*, 1961, **14**, 71.

HILL, J. M. M. & TRIST, E. L. A Consideration of Industrial Accidents as a Means of Withdrawal from the Work Situation. *Hum. Relat.*, 1953, **6**, 357.

HILL, J. M. M. & TRIST, E. L. Changes in Accidents and Other Absences with Length of Service. *Hum. Relat.*, 1955, **8**, 121.

HOEBEL, E. A. *The Law of Primitive Man: a Study in Comparative Legal Dynamics*. Cambridge, Mass.: Harvard Univ. Press, 1954.

JAQUES, E. *The Changing Culture of a Factory*. London: Tavistock Publications, 1951.

JAQUES, E. *Measurement of Responsibility*. London: Tavistock Publications, 1956; Cambridge, Mass., Harvard Univ. Press.

JAQUES, E. *Equitable Payment*. London: Heinemann, 1961.

KATZ, D. & KAHN, R. L. Human Organization and Worker Motivation. In Tripp *et al.* (Eds.), *Industrial Productivity*. Madison, Wisconsin: Industrial Relations Research Association, 1951.

KATZ, D. & SCHANCK, R. C. *Social Psychology*. London: Chapman and Hall, 1938.

LEWIN, K. *Dynamic Theory of Personality*. New York: McGraw-Hill, 1935.

LEWIN, K. *Field Theory in Social Science.* New York: Harper, 1951; London: Tavistock Publications.

LIKERT, R. *New Patterns of Management.* New York: McGraw-Hill, 1961.

LIKERT, R. & HAYES, S. P. (Eds.) *Some Applications of Behavioural Research.* Paris: Unesco, 1957.

LIPPITT, R. *Training in Community Relations.* New York: Harper, 1949.

LIPPITT, R., WATSON, J. & WESTLEY, B. *The Dynamics of Planned Change.* New York: Harcourt, Brace, 1958.

LYMAN, J. Manpower Systems in Large Scale Operations. *Proceedings of the 4th Conference on Research Development in Personnel Management.* University of California, Los Angeles, 1961.

MCGREGOR, D. *The Human Side of Enterprise.* New York: McGraw-Hill, 1960.

MENZIES, I. E. P. A Case-Study in the Functioning of Social Systems as a Defence against Anxiety. *Hum. Relat.*, 1960, **13**, 95.

MERTON, R. K. *Social Theory and Social Structure.* Glencoe, Ill.: Free Press, 1949. (Rev. ed., 1957).

MILLER, E. J. Technology, Territory and Time: The Internal Differentiation of Complex Production Systems. *Hum. Relat.*, 1959, **12**, 243.

PARSONS, T. *The Social System.* Glencoe, Ill.: Free Press, 1951; London: Tavistock Publications.

PEDERSEN-KRAG, G. A Psychoanalytic Approach to Mass Production. *Psychoanal. Quart.*, 1958, **65**, 325.

POLITICAL AND ECONOMIC PLANNING. *Three Case Studies in Automation.* London: P.E.P., 1957.

REID, C. (Chairman). *Coal Mining.* Report of the Technical Advisory Committee (Cmd. 6610). London: H.M.S.O., 1945.

RICE, A. K. *Productivity and Social Organization: The Ahmedabad Experiment.* London: Tavistock Publications, 1958.

RICE, A. K. The Use of Unrecognized Cultural Mechanisms in an Expanding Machine-Shop. *Hum. Relat.*, 1951, **4**, 143.

RICE, A. K. & TRIST, E. L. Institutional and Sub-institutional Determinants of Change in Labour Turnover (The Glacier Project—VIII). *Hum. Relat.*, 1952, **5**, 347.

ROETHLISBERGER, F. J., DIČKSON, W. J., WRIGHT, A. H. *Management and the Worker.* Cambridge, Mass.: Harvard Univ. Press. 1946.

SCOTT, J. F. & LYNTON, R. P. *Three Studies in Management.* London: Routledge & Kegan Paul, 1952.

SELZNICK, P. *Leadership in Administration.* Evanston, Ill.: Row, Peterson, 1957.

SHEPARD, H. A. & BLAKE, R. R. *Changing Behaviour Through Cognitive Change.* Proc. 20th Anniversary Congress of the Soc. of Applied Soc. Anthrop., 1961.

SHEPPARD, W. V. Continuous Longwall Mining. *The Colliery Guardian*, 1949, *178*, 855.

SHEPPARD, W. V. Continuous Longwall Mining: Experiment at Bolsover Colliery. *Colliery Guardian*, 1951, **182**.

STOCK, Dorothy & THELEN, H. A. *Emotional Dynamics and Group Culture.* New York: New York Univ. Press, 1958.

STOUFFER, S. A. *et al. The American Soldier*, Vol. 1. Princeton, N.J.: Princeton Univ. Press, 1949.

TECHNICAL MISSION OF THE NATIONAL COAL BOARD. *The Coal Industry of the U.S.S.R. Part 2.* London: National Coal Board, 1957.

THELEN, H. A. *Dynamics of Groups at Work.* Chicago: University of Chicago Press, 1954.

TOURAINE, Alain. *L'evolution du travail ouvrier aux usines Renault.* Paris: Centre National de la Recherche Scientifique, 1955.

TRIST, E. L. & SOFER, C. *Exploration in Group Relations.* Leicester: Leicester Univ. Press, 1959.

VICKERS, G. *The Undirected Society.* Toronto: Univ. of Toronto Press, 1959; London: Oxford Univ. Press.

WALKER, C. R. & GUEST, R. H. *The Man on the Assembly Line.* Cambridge, Mass.: Harvard Univ. Press, 1952.

WARNER, W. L. & LOW, J. O. *The Social System of the Modern Factory.* New Haven: Yale Univ. Press, 1947.

WESTERLUND, Gunnar. *Group Leadership.* Stockholm: Nordisk Rotogravyr, 1952.

WILLIAMS, R. H. *The Theory of Action in Operations Research and in Social Sciences.* Operations Research Office, T-13, Johns Hopkins University 1950.

WILSON, A. T. M. Some Contrasting Socio-Technical Production Systems. *The Manager*, Dec. 1955.

WOODWARD, J. *Management and Technology.* London: H.M.S.O., 1958.

Index

INDEX